T0122888

Analytical Advances in Quantum and Celestial Mechanics

Separating rapid and slow subsystems

Analytical Advances in Quantum and Celestial Mechanics

Separating rapid and slow subsystems

Eugene Oks
Physics Department, Auburn University, Auburn, AL, USA

IOP Publishing, Bristol, UK

ISBN 978-0-7503-2512-7 (ebook)
ISBN 978-0-7503-2510-3 (print)
ISBN 978-0-7503-2511-0 (mobi)

DOI 10.1088/2053-2563/ab3db0

Version: 20191001

IOP ebooks
ISSN 2053-2563 (online)
ISSN 2054-7315 (print)

British Library Cataloguing-in-Publication Data: A catalogue record for this book is available from the British Library.

Published by IOP Publishing, wholly owned by The Institute of Physics, London

IOP Publishing, Temple Circus, Temple Way, Bristol, BS1 6HG, UK

US Office: IOP Publishing, Inc., 190 North Independence Mall West, Suite 601, Philadelphia, PA 19106, USA

To my muse—the Greek goddess of peace, the freedom from disturbance required for any creative pursuit

Contents

Author biography

Eugene Oks

Eugene Oks received his PhD degree from the Moscow Institute of Physics and Technology, and later the highest degree of Doctor of Sciences from the Institute of General Physics of the Academy of Sciences of the USSR by the decision of the Scientific Council led by the Nobel Prize winner, academician A M Prokhorov. According to the Statute of the Doctor of Sciences degree, this highest degree is awarded only to the most outstanding PhD scientists who founded a new research field of great interest. Oks worked in Moscow (USSR) as the head of a research unit at the Center for Studying Surfaces and Vacuum, then—at the Ruhr University in Bochum (Germany) as an invited professor, and for the last 28 years—at the Physics Department of the Auburn University (USA) in the position of Professor. He conducted research in five areas: atomic and molecular physics, plasma physics, laser physics, nonlinear dynamics, and astrophysics. He founded/co-founded and developed new research fields, such as intra-Stark spectroscopy (new class of nonlinear optical phenomena in plasmas), masing without inversion (advanced schemes for generating/amplifying coherent microwave radiation), and quantum chaos (nonlinear dynamics in the microscopic world). He also developed a large number of advanced spectroscopic methods for diagnosing various laboratory and astrophysical plasmas—the methods that were then used and are used by many experimental groups around the world. He published over 400 papers and 6 books, including the books *Breaking Paradigms in Atomic and Molecular Physics*, *Diagnostics of Laboratory and Astrophysical Plasmas Using Spectral Lineshapes of One-, Two, and Three-Electron Systems*, and *Unexpected Similarities of the Universe with Atomic and Molecular Systems: What a Beautiful World*. He is the Chief Editor of the journal *International Review of Atomic and Molecular Physics*. He is a member of the Editorial Boards of the two other journals: *International Journal of Spectroscopy* and *Open Journal of Microphysics*. He is also a member of the International Program Committees of the two series of conferences: Spectral Line Shapes, as well as Zvenigorod Conference on Plasma Physics and Controlled Fusion.

IOP Publishing

Analytical Advances in Quantum and Celestial Mechanics
Separating rapid and slow subsystems
Eugene Oks

Chapter 1

Introduction

Real physical systems very rarely allow exact analytical solutions of the equations describing the systems. Therefore, mathematicians and theoretical physicists over the last several centuries invented a variety of approximate analytical methods. Especially important are the methods going beyond the perturbation theory, such as, e.g., the method based on the separation of rapid and slow subsystems.

The latter method relates to systems consisting of two subsystems 1 and 2, where the characteristic time of the evolution of subsystem 1 is much smaller than the characteristic time of the evolution of subsystem 2. The two subsystems can interact very strongly, so that their interaction cannot be taken into account by the perturbation theory. In distinction, the method based on the separation of rapid and slow subsystems can provide an approximate analytical solution regardless of the strength of the interaction of the two subsystems.

In studies of nonlinear dynamical systems, the corresponding approach is called the method of averaging and is applied to systems allowing the time-scales separation. The averaging method was developed by Krylov and Bogoliubov and then further developed by Bogoliubov and Mitropolsky, as presented in the book [1] by the latter two authors. A good historical review of their work and of their predecessors' works was presented by Oliveira [2]. The Krylov–Bogoliubov–Mitropolsky's method of averaging and some of its variations were later presented in numerous books, such as, e.g. books [3–7].

A very important contribution to the subject was made by P L Kapitza, the Nobel Prize winner. He was motivated by puzzling experiments with a rigid pendulum, in which the pivot point vibrates in a vertical direction, up and down (later it was called Kapitza's pendulum). The vibrating suspension can cause it to balance stably in an inverted position, with the bob above the suspension point. Kapitza was the first to analyze it in 1951 [8, 9]. He performed experimental studies and provided an analytical explanation of the stability by splitting the motion into 'fast' and 'slow' variables and, most importantly, by introducing the concept of an *effective potential*.

The Kapitza's effective potential describing the 'slow' motion was presented also in the *Mechanics* volume of the Landau–Lifshitz' Course of Theoretical Physics [10]. Here is a brief description of the Kapitza's method.

Let us consider a particle in a static potential $U(x_\alpha)$ and under the force $\boldsymbol{f}(x_\alpha)\cos\omega t$. We seek the solution of the equation of the motion $m\mathrm{d}^2(x_\alpha)/\mathrm{d}t^2 = -\mathrm{d}U/\mathrm{d}x_\alpha + f_\alpha \cos\omega t$ in the form: $x_\alpha(t) = X_\alpha(t) + \xi_\alpha(t)$ and expand the right side in powers of the rapid oscillations ξ_α:

$$\begin{aligned} m\mathrm{d}^2(X_\alpha)/\mathrm{d}t^2 + m\mathrm{d}^2(\xi_\alpha)/\mathrm{d}t^2 = &- \mathrm{d}U/\mathrm{d}X_\alpha - \xi_\alpha \mathrm{d}^2U/\mathrm{d}X_\alpha \mathrm{d}X_\beta \\ &+ [f_\alpha(\mathbf{X}) + \xi_\beta \mathrm{d}f_\alpha/\mathrm{d}X_\beta]\cos\omega t. \end{aligned} \tag{1.1}$$

Here and below, the summation over repeated subscripts is understood.

For the oscillatory terms it is sufficient to write $m\mathrm{d}^2(\xi_\alpha)/\mathrm{d}t^2 = f_\alpha(\mathbf{X})\cos\omega t$, so that $\xi_\alpha = -[f_\alpha/(m\omega^2)]\cos\omega t$. Substituting the latter formula in equation (1.1) and averaging over the period $2\pi/\omega$, one obtains the equation for the averaged motion $\boldsymbol{X}(t)$:

$$m\mathrm{d}^2(X_\alpha)/\mathrm{d}t^2 = -\mathrm{d}U/\mathrm{d}X_\alpha - [1/(2m\omega^2)]f_\beta\, \mathrm{d}f_\alpha/\mathrm{d}X_\beta. \tag{1.2}$$

In the one-dimensional case, equation (1.2) can be interpreted such that the averaged motion of the particle occurs in the following effective potential

$$U_{\mathrm{eff}} = U + f^2/(4m\omega^2). \tag{1.3}$$

Equations (1.2) and (1.3) represent an efficient analytical tool if the force amplitude **f** has gradients. A generalization of Kapitsa's effective potential for a spatially-uniform force amplitude **f**, has been provided by Nadezhdin and Oks [11] and will be presented in chapter 3.

Both the Krylov–Bogoliubov–Mitropolsky method and the Kapitza method are related to classical mechanics and applied to classical systems. In quantum mechanics there is the corresponding method of separating rapid and slow subsystems. It is briefly presented below following, for example, [12].

Let us consider a system consisting of two subsystems 1 and 2 described by the Hamiltonian

$$H(x, \xi) = H_1(x) + V(x, \xi) + H_2(\xi), \tag{1.4}$$

where x and ξ are the coordinates of subsystems 1 and 2, respectively. Here $H_1(x)$ is the Hamiltonian of the isolated subsystem 1, $H_2(x)$ is the Hamiltonian of the isolated subsystem 2, and $V(x, \xi)$ is the interaction between these subsystems: the interaction $V(x, \xi)$ can be strong, so that the perturbation theory would be inapplicable. The characteristic time of the evolution of subsystem 1 is much smaller than the characteristic time of the evolution of subsystem 2. In other words, subsystem 1 is rapid while subsystem 2 is slow.

Let us denote by $\Psi_n(x, \xi)$ and $E_n(\xi)$ the 'instantaneous' eigenfunctions and the eigenvalues, respectively, of the truncated Hamiltonian

$$H_{\mathrm{tr}}(x, \xi) = H_1(x) + V(x, \xi) \tag{1.5}$$

at any fixed value of the coordinate ξ of the slow subsystem. In other words, at this step we 'freeze' the slow subsystem and consider ξ as a parameter, rather than a dynamical variable. So:

$$[H_1(x) + V(x, \xi)]\, \Psi_n(x, \xi) = E_n(\xi)\Psi_n(x, \xi). \tag{1.6}$$

$\Psi_n(x, \xi)$ and $E_n(\xi)$ are assumed to be known for any ξ.

We seek the eigenfunctions of the complete Hamiltonian $H(x, \xi)$ from equation (1.4) in the form

$$\Psi_{nm}(x, \xi) = \Phi_{nm}(\xi)\Psi_n(x, \xi), \tag{1.7}$$

so that

$$[H_1(x) + V(x, \xi) + H_2(\xi)]\Phi_{nm}(\xi)\Psi_n(x, \xi) = E_{nm}(\xi)\Phi_{nm}(\xi)\Psi_n(x, \xi). \tag{1.8}$$

Here, $E_{nm}(\xi)$ are yet unknown eigenvalues of the complete Hamiltonian.

After taking into account equation (1.6), multiplying both parts of equation (1.8) by $\Psi_n^*(x, \xi)$ on the left side, integrating over the coordinate x of the rapid subsystem, and neglecting the action of the operator $H_2(\xi)$ on ξ inside $\Psi_n(x, \xi)$ (i.e., approximating $H_2\Phi\Psi$ by $\Phi H_2\Psi$), we obtain:

$$[H_2(\xi) + E_n(\xi)]\Phi_{nm}(\xi) = E_{nm}(\xi)\Phi_{nm}(\xi). \tag{1.9}$$

Equation (1.9) does not depend on the coordinate x of the rapid subsystem and determines the approximate eigenfunctions $\Phi_{nm}(\xi)$ and eigenvalues $E_{nm}(\xi)$ of the slow subsystem. It is seen that the evolution of the slow subsystem occurs in the effective potential

$$U_{\text{eff}}(\xi) = E_n(\xi). \tag{1.10}$$

In other words, the role of the effective potential for the averaged motion of the slow subsystem is played by the eigenvalues $E_n(\xi)$ of the rapid subsystem.

In atomic and molecular physics, until recently the applications of the method based on the separation of rapid and slow subsystems were limited virtually to only one, though important example: the treatment of the electronic and nuclear motions in molecules, where this method has been known as the Born–Oppenheimer approximation [13] since 1927. To the best of my knowledge, there is no book presenting recent advances in applying this method to other quantum systems. This book is intended to fill this gap. It is devoted primarily to recent advances in applying this method to other quantum systems, such as, e.g., hydrogen atoms in a high-frequency laser field, quantum rotator-dipole in a high-frequency monochromatic field, one-electron Rydberg quasimolecules in a magnetic field, the dynamical Stark broadening of hydrogen spectral lines by plasma ions, the dynamical Stark broadening of hydrogen spectral lines by plasma electrons, and the dynamical Stark broadening of hydrogen-like spectral lines by plasma electrons—see, e.g., review [14].

The book also presents a novel application of the corresponding classical method to some classical systems in general and to three-body systems in celestial mechanics in particular. The latter application results is several particular analytical solutions for the unrestricted three-body problem of celestial mechanics. The term 'unrestricted' means that it is really a three-dimensional motion of the third body: it is not confined in a plane (in distinction to previous analytical solutions). For example, in one of the cases, the orbit of the third body turns out to have a 'corkscrew' shape (see papers [15, 16]).

Last but not least: the book presents also a formalism for the general analytical treatment of quantum systems in a high-frequency field. This formalism is described in appendix A.

As for focusing at the advances in the analytical theory (versus simulations), the following should be noted. Of course, simulations are important as the third powerful research methodology—in addition to theories and experiments: large-scale codes have been created to simulate lots of complicated phenomena. However, first, not all large-scale codes are properly verified and validated, as illustrated by some well-known failures of large-scale codes (see, e.g., [17, 18]). Second, fully-numerical simulations are generally not well-suited for capturing so-called emergent principles and phenomena, such as, e.g., conservation laws and preservation of symmetries, as explained in [17]. Third, as any fully-numerical method, they lack the physical insight.

References

[1] Bogoliubov N N and Mitropolski Y A 1961 *Asymptotic Methods in the Theory of Nonlinear Oscillations* (New York: Gordon and Breach)
[2] Oliveira A R E 2017 *Adv. Histor. Stud.* **6** 40
[3] Vainshtein L A and Vakman D E 1983 *Frequencies Separation in the Theory of Oscillations and Waves* (Moscow: Nauka) in Russian
[4] Guckenheimer J and Holmes P J 1983 *Nonlinear Oscillations, Dynamical Systems, and Bifurcations of Vector Fields* (New York: Springer) ch 4
[5] Verhulst F 1993 *Nonlinear Differential Equations and Dynamical Systems* (New York: Springer) ch 11
[6] Georgescu A 1995 *Asymptotic Treatment of Differential Equations* (London: Chapman & Hall) sections 2.7, 2.8
[7] Sanders J A, Verhulst F and Murdock J 2007 *Averaging Methods in Nonlinear Dynamical Systems Applied Mathematical Sciences* vol 59 (New York: Springer)
[8] Kapitza P L 1951 *Sov. Phys. JETP* **21** 588
[9] Kapitza P L 1951 *Uspekhi Fiz. Nauk* **44** 7
[10] Landau L D and Lifshitz E M 1976 *Mechanics* (Amsterdam: Elsevier) sec. 30
[11] Nadezhdin B B and Oks E 1986 *Sov. Tech. Phys. Lett.* **12** 512
[12] Galitski V, Karnakov B, Kogan V and Galitski V Jr. 2013 *Exploring Quantum Mechanics* (Oxford: Oxford University Press) Problem 8.55
[13] Born M and Oppenheimer J R 1927 *Ann. Phys.* **389** 457
[14] Oks E 2018 *Atoms* **6** 50
[15] Oks E 2015 *Astrophys. J.* **804** 106
[16] Kryukov N and Oks E 2017 *J. Astrophys. Aerospace Technol.* **5** 144
[17] Post D E and Votta L G 2005 *Phys. Today* **58** 35
[18] Oks E 2010 *In 20th Int. Conf. Spectral Line Shapes 2010, AIP Conf. Proc.* **1290** 6

Chapter 2

Quantum hydrogenic atoms or ions in a high-frequency laser field

Because of the continuing advances in developing laser of the far-ultraviolet and x-ray ranges, studies of the behavior of atoms under a high-frequency laser field are of theoretical and practical interest. The 'high-frequency' means that the laser frequency ω is much greater than any of the atomic transition frequencies $\omega_{\mathrm{n'n}}$:

$$\omega \gg \omega_{\mathrm{n'n}}. \tag{2.1}$$

It is well-known [1, 2] that for quantum systems in a monochromatic field it is convenient using the formalism of quasienergy states. The problem of finding such states of the hydrogen-like atom/ion in a linearly-polarized high-frequency laser field was first considered by Ritus [1]. He found quasienergies for the states of the principal quantum number $n = 1$ and $n = 2$, for which the perturbation operator U is diagonal in the basis of the spherical wave functions (i.e., the wave functions of the unperturbed atom in the spherical quantization), as Ritus noted.

In papers [3, 4] there was stated without any proof that for states of $n > 2$, the perturbation operator U couples the substates of l and $l \pm 2$, where l is the orbital momentum quantum number. In paper [4] the study of the quasienergies for $n > 2$ was based on the approximate analogy with the problem of finding energies for the hydrogen molecular ion H_2^+. Based on the energies for H_2^+ from paper [5], for $n > 2$ the author of paper [4] found 'corrections' to the quasienergies from paper [1] due to the coupling of the substates of l and $l \pm 2$. However, in a later paper [6], where the authors used the dependence of the energies for H_2^+ on the internuclear distance R from paper [7], they obtained the result that in the limit $R \to \infty$ coincides with the quasienergies from paper [1], which contradicts paper [4].

Thus, it remained unclear whether the perturbation U couples the substates of l and $l \pm 2$ at fixed n. For answering this question, one should have directly calculated matrix elements of the perturbation between the substates of l and $l \pm 2$ at fixed n.

This was accomplished in paper [8], where the authors demonstrated that these matrix elements are zeros. This meant that in reality, the perturbation operator U does not couple the substates of l and $l \pm 2$ for any n, so that the expression for quasienergies from paper [1] is not limited by $n = 1$ and $n = 2$ (as asserted in paper [1]), but is in fact valid for any n. The diagonality of the perturbation operator U in the basis of the spherical wave functions allowed easily calculating the splitting of *any spectral line* of a hydrogenic atom/ion under the high-frequency laser field. This was the main, fundamental result of paper [8].

Under the condition (2.1), the laser field represents the rapid subsystem, while the hydrogenic atom/ion represents a slow subsystem. The powerful method of separating rapid and slow subsystems allows obtaining accurate analytical results for such systems. The general description of the method can be found in appendix A. As for implementing this method for particular physical systems, there can be some interesting nuances or versions. It is instructive to see how this method was implemented in paper [8], whose results we present below.

The Schrödinger equation for a hydrogen-like atom/ion in a laser field (of the amplitude E_0) described by the vector-potential $\boldsymbol{A}(t) = \boldsymbol{A}_0 \sin \omega t$, where $\boldsymbol{A}_0 = (0, 0, -cE_0/\omega)$, has the following form (here and below the atomic units $\hbar = m_e = e = 1$ are used):

$$
\begin{aligned}
&i\partial\Psi/\partial t = [H_0 + V(t)]\Psi, \quad H_0 = \boldsymbol{p}^2/2 - Z/r + A_0^2/(2c), \\
&V(t) = -(\boldsymbol{A}_0\boldsymbol{p}/c)\sin \omega t - \left[A_0^2/(2c)\right]\cos 2\omega t.
\end{aligned}
\tag{2.2}
$$

Here Z is the nuclear charge and r is the distance of the electron from the nucleus; the notation $\boldsymbol{A}_0\boldsymbol{p}$ stands for the scalar product (also known as the dot-product) of these two vectors. The term $A_0^2/(2c)$ is the average vibrational energy of the free electron in the laser field.

We seek the solution of equation (2.2) in the form

$$
\Psi(t) = \exp[-i\alpha(t)]\Phi, \quad \alpha(t) = [\boldsymbol{A}_0\boldsymbol{p}/(\omega c)]\cos \omega t - \left[A_0^2/(8\omega c^2)\right]\sin 2\omega t. \tag{2.3}
$$

Substituting equation (2.3) in equation (2.2), we obtain the following equation

$$
\begin{aligned}
i\partial\Phi/\partial t &= H_1\Phi, \\
H_1 &= \exp[i\alpha(t)]H_0 \exp[-i\alpha(t)] \\
&= H_0 + i[\alpha, H_0] + (i^2/2)[\alpha,[\alpha, H_0]] + \cdots = H_{1,\text{stat}} + H_{1,\text{osc}}.
\end{aligned}
\tag{2.4}
$$

In equation (2.4), $[\alpha, H_0]$ and $[\alpha, [\alpha, H_0]]$ are commutators; $H_{1,\text{stat}}$ and $H_{1,\text{osc}}$ are the stationary (i.e., averaged over the laser field period $2\pi/\omega$) and oscillatory parts of the Hamiltonian H_1. (It could be instructive to compare the second line of equation (2.1) with the general formula (A.7) from appendix A.) Under the condition (2.1), i.e., since the laser field is the rapid subsystem, the primary contribution to the solution of equation (2.4) originates from the stationary part $H_{1,\text{stat}}$. According to the second line of equation (2.4), $H_{1,\text{stat}}$ can be represented in the following form

$$
\begin{aligned}
H_{1,\text{stat}} &= H_0 + \gamma V_1 + \gamma^2 V_2 + \cdots, \qquad \gamma = E_0^2/(2\omega^2)^2, \\
V_1 &= Z(1 - 3\cos^2\theta)/r^3, \\
V_2 &= 3Z(-3 + 30\cos^2\theta - 35\cos^4\theta)/(4r^5),
\end{aligned} \tag{2.5}
$$

where θ is the polar angle of the atomic electron, the z-axis being parallel to the laser field.

Assuming $\gamma \ll 1$, we will use the perturbation theory for finding the eigenvalues and the eigenfunctions of the Hamiltonian $H_{1,\text{stat}}$. It is important to emphasize the following *counterintuitive* fact: since the unperturbed system is degenerate, then according to paper [9] the linear (with respect to γ) corrections to the eigenfunctions will originate not only from the term γV_1, but also from the term $\gamma^2 V_2$.

It is easy to see that the radial part of the matrix element $\langle nlm|V_1|nl'm\rangle$, where $l' = l - 2$, reduces to the following type of the integral:

$$
J = \int_0^\infty z^r \exp(-z) Q_{k+s}{}^r(z) Q_k{}^p(z) dz, \quad (s > 0), \tag{2.6}
$$

where $Q_n{}^m(z)$ are the Laguerre polynomials. According to the textbook [10], for $\langle nlm|V_1|nl'm\rangle$ with $l' = l - 2$ one gets $J = 0$, so that $<nlm|V_1|nl'm> = 0$. This means that the spherical eigenfunctions φ_{nlm} of the unperturbed Hamiltonian H_0 turn out to be the correct eigenfunctions of the zeroth order of the truncated perturbed Hamiltonian $H_0 + \gamma V_1$. Therefore, according to paper [9], the eigenvalues $F_{n\lambda}$ and the eigenfunctions $\chi_{n\lambda}$ of the Hamiltonian $H_{1,\text{stat}}$ within the accuracy of the terms $\sim\gamma$ are expressed as follows

$$
F_{n\lambda} = E_n{}^{(0)} + \gamma\langle n\lambda|V_1|n\lambda\rangle, \qquad E_n{}^{(0)} = -Z^2/(2n^2) + E_0{}^2/(2\omega)^2, \tag{2.7}
$$

$$
\begin{aligned}
\chi_{n\lambda} = \varphi_{n\lambda} &+ \gamma\sum_j \langle j|V_1|n\lambda\rangle\varphi_j/(E_n{}^{(0)} - E_j{}^{(0)}) \\
&+ \gamma\sum_{\mu\neq\lambda}\Bigg\{\sum_j \langle n\mu|V_1|j\rangle\langle j|V_1|n\lambda\rangle\varphi_{n\mu}/[(E_n{}^{(0)} - E_j{}^{(0)}) \\
&\times(\langle n\lambda|V_1|n\lambda\rangle - \langle n\mu|V_1|n\mu\rangle)] \\
&+ \langle n\mu|V_1|n\lambda\rangle\varphi_{n\mu}/(\langle n\lambda|V_1|n\lambda\rangle - \langle n\mu|V_1|n\mu\rangle)\Bigg\}.
\end{aligned} \tag{2.8}
$$

In equation (2.8), $\lambda = (l, m)$, $\mu = (l', m')$, $\mathrm{j} = (n', l', m')$, $n' \neq n$.

Substituting V_1 from equation (2.5) in equation (2.8), we obtain the following for $l > 0$:

$$
\begin{aligned}
F_{nlm} = &-Z^2/(2n^2) + E_0{}^2/(2\omega)^2 + (Z^4 E_0{}^2/\omega^4) \\
&\times[3m^2 - l(l+1)]/[n^3(2l+3)(l+1)(2l+1)l(2l-1)].
\end{aligned} \tag{2.9}
$$

For finding F_{nlm} (i.e., for $l = 0$), instead of the formula for V_1 from equation (2.5), we use the following expression:

$$V_1 = Z[1 - 3\cos^2\theta + \varepsilon(4\cos^2\theta - 1)]r^{\varepsilon-3}, \quad |\varepsilon| \ll 1. \tag{2.10}$$

The expression for V_1 from equation (2.10) corresponds to the quasi-Coulomb nuclear potential $-Zr^{\varepsilon-1}$. This trick allows removing the uncertainty that would otherwise arise while calculating matrix elements of the operator V_1 in the basis of the eigenfunctions φ_{n00}. For completeness we note that a similar uncertainty arises while calculating matrix elements $\langle nlm|V_2|\, n', l', m\,\rangle$; in this case one should use the expression

$$V_2 = 3Z\{-3 + 30\cos^2\theta - 35\cos^4\theta + \varepsilon[4 - 46\cos^2\theta + (176/3)\cos^4\theta]\}/(4r^{5-\varepsilon}), \tag{2.11}$$

corresponding to the quasi-Coulomb nuclear potential $-Zr^{\varepsilon-1}$.

After calculating the necessary matrix elements by using the potential V_1 from equation (2.10) and then setting $\varepsilon = 0$, we obtain the following result for the energy F_{n00} (i.e., for $l = 0$):

$$F_{n00} = -Z^2/(2n^2) + E_0{}^2/(2\omega)^2 + Z^4E_0{}^2/(3n^3\omega^4). \tag{2.12}$$

It is worth noting that the above rigorously calculated result for F_{n00} can be also formally obtained from the right side of equation (2.9) in the following three steps:

1) to set $m = 0$;
2) to cancel out $l(l + 1)$ in the numerator and denominator;
3) to set $l = 0$.

Thus, indeed the expression for quasienergies from paper [1] is not limited by $n = 1$ and $n = 2$ (as asserted in paper [1]). The above proves that they are actually applicable for any n.

For the validity of the above results it is necessary that the characteristic value of the splitting of the energy level of the principal quantum number n, determined by equations (2.9), (2.12), significantly exceeded the energy shift Δ_{nlm} caused by the term $H_{1,\text{osc}}$ in equation (2.4). By limiting ourselves by the term in $H_{1,\text{osc}}$, containing the small parameter γ in the lowest degree (i.e., by the term proportional to $\gamma^{1/2}/\omega$), in the high-frequency limit defined by equation (2.1) we obtain the following relation (see, e.g., book [11])

$$\Delta_{nlm} = -ZE_0{}^{(2)}/(2\omega^6)\sum_{n'\neq n} |\langle nlm|r^{-2}\cos\theta|n'lm\rangle|^2\left(E_n{}^{(0)} - E_{n'}{}^{(0)}\right), \tag{2.13}$$

that serves for finding the lower limit of validity with respect to the laser frequency ω.

The above results were obtained for a *linearly-polarized* high-frequency laser field. A more general case where the high-frequency laser field is *elliptically-polarized* was considered in paper [12]. The vector-potential of the laser electric field was chosen in the form

$$A(t) = A_0(1 + \zeta^2)(\cos\omega t, \zeta\sin\omega t, 0), \quad A_0 = -cE_0/\omega, \tag{2.14}$$

where ζ is the ellipticity degree. For the quasi-Coulomb nuclear potential $-Zr^{\varepsilon-1}$, the analog of the γV_1 from equation (2.5), denoted below as V, now can be represented as follows

$$V = [Z/(2c^2)](\varepsilon + 1)\langle \mathbf{B}^2 \rangle / r^{3+\varepsilon}, \quad |\varepsilon| \ll 1. \tag{2.15}$$

In equation (2.15), vector $\mathbf{B}(t)$ is the solution of the equation

$$\mathrm{d}\mathbf{B}(t)/\mathrm{d}t = \mathcal{A}(t) \tag{2.16}$$

and has the zero time average; the notation <...> stands for the average over the period of the laser field.

Substituting equation (2.14) into equation (2.15) and using the spherical coordinates, paper [12] found the following:

$$\begin{aligned} V(\zeta^2) = -&\left[ZE_0{}^2/(8\omega^4 r^{3+\varepsilon})\right] \\ &\times \{(1 - 3\cos^2\theta) - 2\varepsilon\cos 2\theta + [3(1 - \zeta^2)/(1 + \zeta^2)]\sin^2\theta \\ &\cos 2\varphi\}. \end{aligned} \tag{2.17}$$

In paper [13], it was shown that

$$\int_0^\infty r^{-s} R_{nl}(r) R_{nl'}(r) r^2 \,\mathrm{d}r = 0, \quad s = 2, 3, 4, \ldots, l - l' + 1, \tag{2.18}$$

where $R_{nl}(r)$ are the radial wave functions of the hydrogenic atom/ion. Therefore, the matrix elements of the operator $V(\zeta^2)$ satisfy the following relation:

$$\langle nlm| V(\zeta^2) |nl'm'\rangle = 0, \quad l' = l \pm 2. \tag{2.19}$$

For a particular case of the *circular polarization* ($\zeta^2 = 1$), the term containing cos 2φ in equation (2.17) is absent. Therefore, for the case of the *circular polarization*, due to the relation (2.19), the spherical eigenfunctions φ_{nlm} are the correct eigenfunctions of the zero order for the perturbed Hamiltonian (as was the case for the linear polarization of the laser field). The following energy eigenvalues correspond to these eigenfunctions:

$$F_{nlm} = -Z^2/(2n^2) + E_0{}^2/(2\omega)^2 + \langle nlm| V |nlm\rangle, \tag{2.20}$$

$$\begin{aligned} \langle nlm| V |nlm\rangle \; = \; -&\left[Z^4 E_0{}^2/(2\omega^4)\right] \\ &\times \; [3m^2 - l(l + 1)]/[n^3(2l + 3)(l + 1)(2l + 1)l(2l - 1)], \\ l \; > \; 0,& \end{aligned} \tag{2.21}$$

$$\langle nlm| V |nlm\rangle = -Z^4 E_0{}^2/(6n^3\omega^4), \quad l = 0. \tag{2.22}$$

(In equation (2.22), we corrected a misprint from the corresponding expression in equation (2.10) from paper [12].) Just as in the case of the linear polarization of the laser field, the result presented in equation (2.22) can be obtained from the right side of equation (2.21) in the following three steps:

1) to set $m = 0$;
2) to cancel out $l(l + 1)$ in the numerator and denominator;
3) to set $l = 0$.

So, the expression for quasienergies from paper [1] for the circular polarization of the laser field is not limited by $n = 1$ and $n = 2$ (as asserted in paper [1]). The above proves that they are actually applicable for any n—just as it was in the case of the linear polarization of the laser field.

Now we are coming back to the situation, where the ellipticity degree of the laser field is arbitrary. When $\zeta^2 \neq 1$, the term containing $\cos 2\varphi$ in equation (2.17) will couple the states φ_{nlm} and $\varphi_{nlm'}$ ($l > 0$, $m' = m - 2$). However, the state φ_{n00} would not be coupled by this term and the state φ_{n10} would not be coupled by this term either. Therefore, the energy eigenvalues for these two states do not depend on the ellipticity degree ζ and are still given by equations (2.21), (2.22).

For finding the eigenvalues of the energy for the other states in the general case of $\zeta^2 \neq 1$, one should solve the corresponding secular equation. For the states φ_{nl1} and φ_{nl-1} ($l = 1, 2$), the secular equation is a quadratic one, yielding the following two energy eigenvalues F_{ns} ($s = 1, 2$):

$$\begin{aligned} F_{ns} = &- Z^2/(2n^2) + E_0{}^2/(2\omega)^2 \\ &+ \left\{ Z^4 E_0{}^2/[2\omega^4 n^3(2l + 3)(2l + 1)(2l - 1)] \right\} \\ &\times [(l^2 + l - 3)/(l + 1) + (-1)^{s+1}(3/2)(1 - \zeta^2)/(1 + \zeta^2)], \end{aligned} \tag{2.23}$$

where $l = 1, 2$; $s = 1, 2$. The corresponding eigenfunctions are

$$\varphi_{ns} = 2^{-1/2}[(-1)^{s+1}\varphi_{nl1} + \varphi_{nl-1}], \quad l = 1, 2; s = 1, 2, \tag{2.24}$$

Figure 2.1 shows the scaled splitting $S = (F_{n1} - F_{n2})[2\omega^4 n^3(2l + 3)(2l + 1)(2l - 1)]/Z^4 E_0{}^2$ versus the ellipticity degree for the states of $l = 1, 2$. It is seen that as the

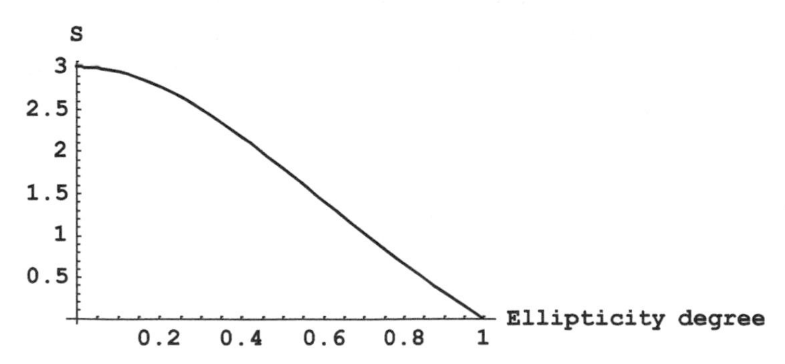

Figure 2.1. The scaled splitting $S = (F_{n1} - F_{n2})\,[2\omega^4 n^3(2l + 3)(2l + 1)(2l - 1)]/Z^4 E_0{}^2$ versus the ellipticity degree for the states of $l = 1, 2$. The energies $F_{n1} - F_{n2}$ are given by equation (2.23).

ellipticity degree ζ increases, the splitting S decreases. For the circular polarization ($\zeta^2 = 1$) the splitting S vanishes.

For the states φ_{nl0}, φ_{nl2}, and φ_{nl-2} ($l = 2, 3$), the secular equation is a cubic one, yielding the following three energy eigenvalues F_{np} ($p = 1, 2, 3$)

$$\begin{aligned} F_{n_1} &= -Z^2/(2n^2) + E_0{}^2/(2\omega)^2 + V_{22}, \\ F_{np} &= -Z^2/(2n^2) + E_0{}^2/(2\omega)^2 + (1/2) \\ &\quad \times \{V_{11} + V_{22} + (-1)^p[(V_{11} - V_{22})^2 + 8V_{12}{}^2]^{1/2},\ p = 2, 3, \end{aligned} \tag{2.25}$$

where

$$V_{11} = Z^4E_0{}^2/[(2\omega^4n^3(2l + 3)(2l + 1)(2l - 1)], \tag{2.26}$$

$$V_{22} = -Z^4E_0{}^2[12 - l(l + 1)]/[(2\omega^4n^3(2l + 3)(2l + 1)(2l - 1)l(l + 1)], \tag{2.27}$$

$$\begin{aligned} V_{12} &= \left[3Z^4E_0{}^2/(4\omega^4)\right][(1 - \zeta^2)/(1 + \zeta^2)] \\ &\quad \times \{(l - 1)(l + 2)/[l(l + 1)]\}/[n^3(2l + 3)(2l + 1)(2l - 1)]. \end{aligned} \tag{2.28}$$

The corresponding eigenfunctions are as follows:

$$\begin{aligned} \varphi_{n1} &= 2^{-1/2}[(-1)^{s+1}\varphi_{nl2} - \varphi_{nl-2}], \\ \varphi_{np} &= a(p)\varphi_{nl0} + b(p)(\varphi_{nl2} + \varphi_{nl-2}), \qquad p = 2, 3, \end{aligned} \tag{2.29}$$

where

$$\begin{aligned} a(p) &= 2^{-1/2}\{1 + (-1)^{p+1}(V_{22} - V_{11})/[(V_{11} - V_{22})^2 + 8V_{12}{}^2]^{1/2}\}^{1/2}, \\ b(p) &= [(-1)^p/2]\{1 + (-1)^p(V_{22} - V_{11})/[(V_{11} - V_{22})^2 + 8V_{12}{}^2]^{1/2}\}^{1/2}. \end{aligned} \tag{2.30}$$

In summary, expressions (2.9), (2.12), (2.20)–(2.30) determine the splitting of hydrogenic spectral lines in the high-frequency laser field. The intensities of the split components can be calculated by using the well-known expressions (e.g., from book [14]) for the matrix elements $|\langle nlm|r^{-2}\cos\theta|n'l'm\rangle|^2$ in the spherical coordinates. In particular, the allowance of the terms $\sim\gamma$ in equation (2.8) would lead to the appearance of the forbidden components (of intensities $\sim\gamma^2$) in the spectra of hydrogenic lines. In general, the observation of the splitting of the spectrum of a hydrogenic line in two different polarizations, allows determining the following three physical quantities:

1) the polarization of the vector-potential $\mathbf{A}(t)$;
2) the ellipticity degree ζ;
3) the amplitude E_0 of the laser field (for the known frequency ω).

References

[1] Ritus V I 1966 *Sov. Phys. JETP* **24** 1041
[2] Zeldovich Y B 1966 *Sov. Phys. JETP* **24** 1006
[3] O'Connell R F 1975 *Phys. Rev.* A **12** 1132

[4] Savukinas A Y 1977 *Litovskiy Fiz. Sbornik* **17** 729
[5] Bates O R, Ledsham K and Stewart A L 1953 *Phil. Trans. Roy. Soc.* **246** 215
[6] Vaitekunas P P and Savukinas A Y 1983 *Opt. Spectrosc. (USSR)* **54** 17
[7] Abramov O I and Slavyanov S Y 1978 *J. Phys.* B **11** 2229
[8] Gavrilenko V P, Oks E and Radchik A V 1985 *Opt. Spectrosc. (USSR)* **59** 411
[9] Sholin G V 1969 *Opt. Spectrosc. (USSR)* **26** 275
[10] Fok V A 1976 *Fundamentals of Quantum Mechanics* (Moscow: Mir) part 2, section 5.4
[11] Sobelman I I 1973 *Introduction to the Theory of Atomic Spectra* (Oxford: Pergamon)
[12] Gavrilenko V P 1985 *Sov. Phys. JETP* **63** 500
[13] Pasternack S and Sternheimer J 1962 *J. Math. Phys.* **3** 1280
[14] Bethe H A and Salpeter E E 1957 *Quantum Mechanics of One- and Two-Electron Atoms* (Berlin: Springer)

Chapter 3

Classical (Rydberg) hydrogen atoms in a high-frequency laser field: celestial analogies

We consider the situation where the laser frequency ω is much greater than the Kepler frequency $\omega_K = m_e e^4/(n^3\hbar^3)$ of the highly-excited (Rydberg) hydrogen atom:

$$\omega \gg \omega_K. \tag{3.1}$$

Here, m_e and e are the electron mass and charge, respectively; $n \gg 1$ is the principal quantum number. Under the condition (3.1), the laser field constitutes the fast subsystem, while the Rydberg atom constitutes the slow subsystem.

The Kapitsa method of splitting the motion into 'fast' and 'slow' variables and introducing the concept of an *effective potential* for the slow subsystem, in its standard form [1–3] is not applicable for the case of a spatially-uniform amplitude **f** of the oscillatory force $\mathbf{f}(x_\alpha)\cos\omega t$, as mentioned in chapter 1. A generalization of Kapitsa's effective potential for a spatially-uniform force amplitude **f**, has been provided by Nadezhdin and Oks [4] and is presented below.

As in chapter 1, we consider a particle in a static potential $U(x_\alpha)$ and under the force $\mathbf{f}(x_\alpha)\cos\omega t$. We seek the solution of the equation of the motion

$$m\mathrm{d}^2(x_\alpha)/\mathrm{d}t^2 = -\mathrm{d}U/\mathrm{d}x_\alpha + f_\alpha \cos\omega t \tag{3.2}$$

in the form

$$x_\alpha(t) = X_\alpha(t) + \xi_\alpha(t) \tag{3.3}$$

and expand the right side in powers of the rapid oscillations ξ_α:

$$\begin{aligned} md^2(X_\alpha)/\mathrm{d}t^2 + md^2(\xi_\alpha)/\mathrm{d}t^2 = &-\mathrm{d}U/\mathrm{d}X_\alpha - \xi_\alpha\,\mathrm{d}^2U/\mathrm{d}X_\alpha\mathrm{d}X_\beta \\ &-(1/2)\xi_\beta\xi_\gamma\mathrm{d}^3U/\mathrm{d}X_\alpha\mathrm{d}X_\beta\mathrm{d}X_\gamma \\ &+[f_\alpha(\mathbf{X}) + \xi_\beta\mathrm{d}f_\alpha/\mathrm{d}X_\beta]\cos\omega t. \end{aligned} \tag{3.4}$$

Here and below, the summation over repeated subscripts is understood. In distinction to equation (1.1) from chapter 1, here the term—(1/2) $\xi_\beta\xi_\gamma$ $d^3U/dX_\alpha dX_\beta dX_\gamma$ is taken into account.

For the oscillatory terms it is sufficient to write

$$md^2(\xi_\alpha)/dt^2 = f_\alpha(\mathbf{X})\cos\omega t, \tag{3.5}$$

so that

$$\xi_\alpha = -[f_\alpha/(m\omega^2)]\cos\omega t. \tag{3.6}$$

Substituting equation (3.6) in equation (3.4) and averaging over the period $2\pi/\omega$, we obtain the equation for the averaged motion $\mathbf{X}(t)$:

$$\begin{aligned} md^2(X_\alpha)/dt^2 = &-dU/dX_\alpha \\ &-[1/(4m^2\omega^4)]f_\beta f_\gamma d^3U/dX_\alpha dX_\beta dX_\gamma \\ &-[1/(2m\omega^2)]f_\beta df_\alpha/dX_\beta. \end{aligned} \tag{3.7}$$

For a spatially-uniform force $\mathbf{f}$, the term proportional to $1/\omega^2$ in equation (3.7) vanishes, so that it becomes important to take into account the term proportional to $1/\omega^4$. As a result, the equation for the averaged motion takes the form:

$$md^2(X_\alpha)/dt^2 = -d/dX_\alpha\{U + [f_\beta f_\gamma/(4m^2\omega^4)]d^2U/dX_\beta dX_\gamma\}. \tag{3.8}$$

Thus, in this situation we deal with the effective potential

$$U_{\text{eff}} = U + [f_\beta f_\gamma/(4m^2\omega^4)]d^2U/dX_\beta dX_\gamma. \tag{3.9}$$

Oks *et al* [5] applied the effective potential from equation (3.9) to the case of a Rydberg hydrogen atom in an *elliptically-polarized* high-frequency laser field:

$$\mathbf{E}(t) = \mathbf{e}_x\varepsilon_0\cos\omega t + \mathbf{e}_y\mu\varepsilon_0\sin\omega t, \tag{3.10}$$

where μ is the ellipticity degree. The peak field ε_0 in expression (3.10) is connected with the time-average of the electric field as follows:

$$\langle E^2(t)\rangle = \left\langle\left\{\varepsilon_0^2\cos^2\omega t + \mu^2\varepsilon_0^2\sin^2\omega t\right\}\right\rangle = \varepsilon_0^2(1+\mu^2)/2. \tag{3.11}$$

On the other hand, one can define an effective amplitude E_0 through

$$\langle E^2(t)\rangle = E_0^2/2. \tag{3.12}$$

By equating the right sides of equations (3.11) and (3.12), one obtains the following relation between E_0 and ε_0

$$\varepsilon_0 = E_0/(1+\mu^2)^{1/2}. \tag{3.13}$$

By applying the effective potential from equation (3.9) to the case of the high-frequency elliptically-polarized laser field from equation (3.10), Oks *et al* [5] found the following effective potential

$$U_{\text{eff}} = -e^2/r + (\gamma/r^3)\{(1 + \mu^2) - 3\sin^2\theta[1 - (1 - \mu^2)\sin^2\varphi]\}, \quad (3.14)$$

where

$$\gamma = \left(e^4\varepsilon_0^2\right)/\left(4m^2\omega^4\right), \quad (3.15)$$

θ is the polar angle, and φ is the azimuthal angle of the radius vector of the electron (the z-axis is chosen to be perpendicular to the polarization plane).

In the particular case where the high-frequency laser field is *circularly polarized* ($\mu = 1$), the effective potential from equation (3.14) simplifies to:

$$U_{\text{eff}}(r, \theta) = -e^2/r + (\gamma/r^3)(3\cos^2\theta - 1). \quad (3.16)$$

For this case of the circular polarization, by using equation (3.13) with $\mu = 1$, formula (3.15) for γ can be also expressed through the time-average square of the laser field as follows:

$$\gamma = \left(e^4E_0^2\right)/\left(8m^2\omega^4\right). \quad (3.17)$$

The effective potential given in equation (3.16) has the following remarkable feature: it is identical to the potential of a satellite orbiting a prolate planet. The motion of this satellite has been completely investigated in celestial mechanics (see, e.g., book [6], section 10.4). It turns out that not only does the ellipse precess in its plane with some frequency $f_e \ll \Omega$, but the plane of orbit precesses as well with frequency $f_p \sim f_e$.

The precessions above are found via canonical perturbation theory and by employing action-angle formulation (see, e.g., book [7]). In this situation the effective potential from equation (3.1) can be represented in the form of the following perturbation Hamiltonian

$$\Delta H = C(3\cos^2\theta - 1)/r^3, \quad C = k^2(I_3 - I_1)/(2M_0), \quad k = GM_0m, \quad (3.18)$$

where m is the mass of the satellite, M_0 is the mass of the a prolate planet mass, and I_3 and I_1 are the principal moments of inertia, I_1 being the moment of inertia with respect to the axis of symmetry. The time average of the perturbation Hamiltonian is

$$\langle\Delta H\rangle = [m^2k^2(I_3 - I_1)/(2M_0M^3\tau)]\int_0^{2\pi}(1 + \varepsilon\cos\Psi)(3\cos^2\theta - 1)\mathrm{d}\Psi, \quad (3.19)$$

where ε is the eccentricity of the satellite orbit, M is the magnitude of the total angular momentum, τ is the period, and Ψ is the angle of the radius-vector in the orbital plane relative to the periapsis. We use the relation

$$3\cos^2\theta - 1 = [(1/2) - (3/2)\cos^2 i)] - \{(3/2)\sin^2 i\cos[2(\Psi + \omega)]\}, \quad (3.20)$$

where i is the angle between the unperturbed plane of orbit and the equatorial plane of the planet (the angle of inclination). Employing equation (3.20) we find:

$$\langle \Delta H \rangle = \pi m^2 k^2 (I_3 - I_1)(1 - 3\cos^2 i)/(2M_0 M^3 \tau). \tag{3.21}$$

We relate the variable i to the action angle variables J_1 and J_2 as

$$J_1/J_2 = \cos i, \quad J_1 = J_\varphi = 2\pi M_z, \tag{3.22}$$

where M_z is the constant value of angular momentum about the polar axis,

$$J_2 = J_\varphi + J_\theta = 2\pi M. \tag{3.23}$$

Here M is the magnitude of the total angular momentum, and

$$J_3 = \pi k[(2m/(-E)]^{1/2}, \tag{3.24}$$

where k is the constant from the expression for the unperturbed gravitational potential

$$V(r) = -k/r \tag{3.25}$$

and corresponds to e^2 in the atomic problem.

The orbit undergoes two precessions, as follows. Because of the smallness of the perturbation, the precession of the orbital plane around the polar axis shows up as a secular change in Ω (see figure 3.1):

$$\langle \mathrm{d}\Omega/\mathrm{d}t \rangle \tau/(2\pi) = [\partial(\Delta H)/\partial J_1]\tau/(2\pi) = [\tau/(2\pi M)]\partial(\Delta H)/\partial(\cos i). \tag{3.26}$$

Therefore, it follows that the frequency of precession of the orbital plane around the polar axis is given by

$$f_p = -(3/2)(I_3 - I_1)\cos i/[M_0 a^2 (1 - \varepsilon^2)^2], \tag{3.27}$$

where we have made use of the expression

$$M^2 = mka(1 - \varepsilon^2) \tag{3.28}$$

(here a is the semi-major axis).

The second one is the precession of the periapsis of the elliptical orbit in the plane of orbit—see figure 3.2. It is given by

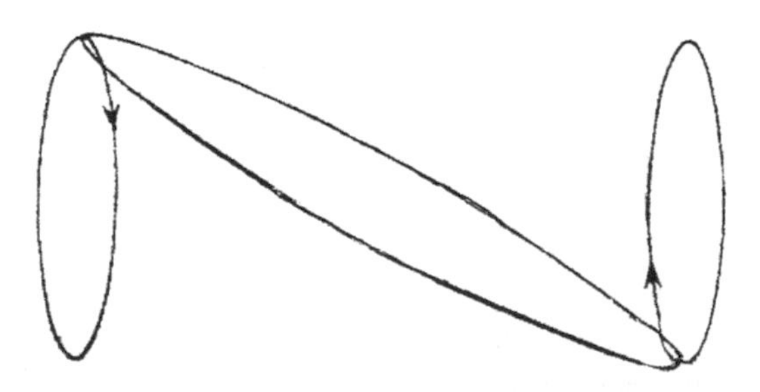

Figure 3.1. Precession of the orbital plane around the polar axis.

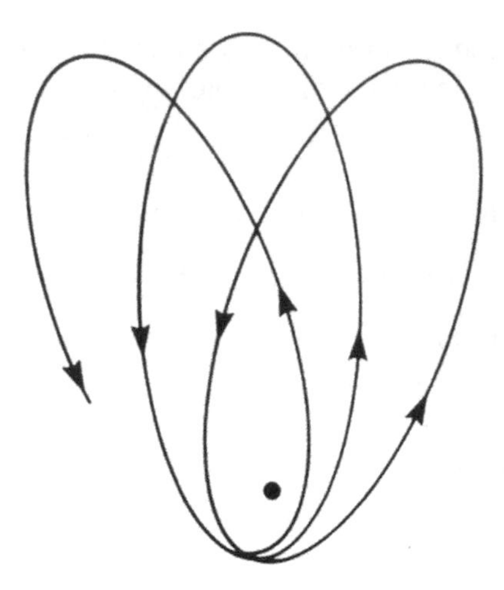

Figure 3.2. Precession of the periapsis of the elliptical orbit in the plane of the orbit.

$$\langle d\omega/dt\rangle\tau/(2\pi) \equiv [\partial(\Delta H)/\partial J_2]\tau/(2\pi) = [\tau/(2\pi)]\partial(\Delta H)/\partial M. \quad (3.29)$$

Upon taking the derivatives, we find

$$f_e = 3(I_3 - I_1)(5\cos^2 i - 1)/[4M_0a^2(1 - \varepsilon^2)^2]. \quad (3.30)$$

The averaged Hamiltonian $\langle \Delta H\rangle$ from equation (3.21) does not depend on the angular variables of the spherical coordinate system. Consequently, the action variables J_1, J_2, J_3 are conserved with respect to the averaged motion. The major semi-axis a and the eccentricity ε_0 of the orbit are the following functions of J_1, J_2, J_3

$$a = J_3^2/(4\pi^2mk), \quad \varepsilon = [(1 - (J_2/J_3)^2)]^{½}. \quad (3.31)$$

Therefore, there is no secular change in either a or ε. Physically this means that the shape and size of this precessing ellipse, when averaged over the orbital period, will not change in time.

The approximate conservation of the shape and size of the precessing ellipse means that the square of the angular momentum $\mathbf{M}^2$ is an approximately conserved quantity (to within the accuracy of terms ~E_0^2). Physically this means that the averaged *system has a higher symmetry than the geometrical symmetry*, which was axial symmetry. In other words, *the system has an algebraic symmetry which is spherical.*

Thus not only does the unperturbed Hamiltonian H_0 commute with M^2 and M_z but the perturbed Hamiltonian $H_0 + V$ commutes (approximately) with M^2 and M_z. This justifies the employment of the non-degenerate classical perturbation theory in the previous section.

This also means that in the quantum treatment of the corresponding atomic problem, the perturbation V is diagonal (in the same approximation) in the basis of the spherical wave functions. Consequently, due to the conservation of $\mathbf{M}^2$, the perturbed energies can be found as mean values of the perturbation over the unperturbed motion. Following paper [5], we find the quantum corrections to the energy levels in the quasiclassical formalism as follows.

The perturbation term in the effective potential is

$$\Delta U = [\gamma / r^3][3\cos^2\theta - 1]. \tag{3.32}$$

We average this expression over the period of the unperturbed motion

$$\langle \Delta U \rangle = [\gamma/\tau] \int_0^{2\pi} [3\cos^2\theta - 1] r^{-3} \mathrm{d}t. \tag{3.33}$$

The angular momentum can be written as $M = mr^2 \mathrm{d}\Psi/\mathrm{d}t$, where Ψ is the angle of the radius vector in the orbital plane relative to the periapsis. Substituting the expression

$$\mathrm{d}t = mr^2 M^{-1} \mathrm{d}\Psi \tag{3.34}$$

into equation (3.33) for the effective potential, we arrive at the following

$$\langle \Delta U \rangle = [\gamma m/(M\tau)] \int_0^{2\pi} (3\cos^2\theta - 1) r^{-1} \mathrm{d}\Psi. \tag{3.35}$$

Using the relations

$$\cos\theta = \mathrm{sin}i\cos\Psi, \quad r^{-1} = (1 + \varepsilon\cos\Psi)p^{-1}, \quad p = M^2(mk)^{-1}, \tag{3.36}$$

where $k = e^2$, we find

$$\begin{aligned} \langle \Delta U \rangle &= [\gamma m^2/(M^3\tau)] \int_0^{2\pi} (3\sin^2 i \cos^2\Psi - 1)(1 + \varepsilon\cos\Psi) \mathrm{d}\Psi \\ &= [(\gamma m^2 e^2)/(M^3\tau)] 2\pi [(3/2)\sin^2 i - 1]. \end{aligned} \tag{3.37}$$

Substituting the expressions for the period

$$\tau = 2\pi n^{3}{\_}^{3}/(me^{4}) \tag{3.38}$$

into equation (3.37), we obtain:

$$\langle \Delta U \rangle = [\gamma m^3 e^6][(3/2)\sin^2 i - 1]/[M^3 \hbar^3 n^3]. \tag{3.39}$$

Transforming

$$[(3/2)\sin^2 i - 1] = (1/2)\{[3(1 - \cos^2 i) - 2]\}, \quad \mathrm{cos}i = M_z/M, \tag{3.40}$$

and replacing

$$M = \hbar(l + 1/2), \quad M_z = m_0\hbar \tag{3.41}$$

in equation (3.37), we finally obtain the corrections to the energy levels:

$$E^1{}_{nlm} \equiv \langle \Delta U \rangle = \left[\left(e^{10} m E_0{}^2\right)\left(8\hbar^6 n^3 \omega^4\right)^{-1}\right]\left\{\left[(l + 1/2)^2 - 3m_0{}^2\right]\left[(l + 1/2)^{-5}\right]\right\}. \tag{3.42}$$

We emphasize again that equation (3.42) is valid for $n,l \gg 1$. Hence, the energy levels are split and the degeneracy is partially removed. Because 'm_0', the magnetic quantum number, is squared, terms where the magnitude of m_0 is the same, but the sign differs, will remain degenerate.

Now, following paper [5], we present the study of the case where the ellipticity degree ξ is arbitrary, but the Keplerian plane and the plane of electromagnetic radiation are coincident. This means that $M_z = M$, and that the polar angle of the radius-vector does not vary in the course of time (i.e., $\theta = \pi/2$). In this case, the perturbed motion is really only two-dimensional because all of the forces acting on the particle are confined to the plane. Our effective potential becomes

$$U_{\text{eff}} = -e^2/r + (\gamma/r^3)[(1 - 2\xi^2) - 3(1 - \xi^2)\cos\varphi]. \tag{3.43}$$

This problem is equivalent to a problem of celestial mechanics in which a satellite revolves in an equatorial orbit about a slightly non-spherical planet. For this case the plane of orbit does not change its orientation in the course of time. Therefore, the only precession that might occur is the precession of the periapsis of the ellipse—similar to that depicted in figure 3.2.

It turns out that M_z is an approximately conserved quantity within the accuracy of terms $\sim E_0^2$. In this case, the system has a dynamical symmetry, which is axial; geometrically there was originally no symmetry at all. Consequently, we simply average this U_{eff} over the unperturbed motion to find the classical perturbed motion and the quasiclassical corrections to the energy levels—in distinction to the case of arbitrary orientations of the polarization and Keplerian planes presented in paper [5], in its appendix.

We can see that

$$\langle \Delta U \rangle = [m/(M\tau)] \int_0^{2\pi} [\gamma(1 - 2\xi^2) - 3\gamma(1 - \xi^2)\cos^2\varphi] r^{-3} \mathrm{d}\varphi, \tag{3.44}$$

where we have made use of equations (3.34) and (3.35). Calculating this integral we obtain

$$\langle \Delta U \rangle = -(\pi m^2 e^2 \gamma / M 3\tau)(1 + \xi^2). \tag{3.45}$$

Recalling the expression for the secular precession of the periapsis in the plane of the orbit

$$[\langle \mathrm{d}\omega/\mathrm{d}t \rangle \tau/(2\pi)] = [\tau/(2\pi)](\partial \langle \Delta U \rangle / \partial M), \tag{3.46}$$

we immediately see that the precession frequency is

$$f_e = [3m^2 e^2 \gamma (1 + \xi^2)]/(2M^4). \tag{3.47}$$

For the same reason as in the case of circular polarization, there is no secular change in either the size a of the major semi-axis of the ellipse or in the eccentricity ε_0. Again, physically this means that the shape and size of this precessing ellipse when averaged over the orbital period will not change in time.

To find quasi-classical energy corrections, we substitute into equation (3.45) expressions (3.36), (3.38), and finally obtain:

$$E^{(1)}_{nlm} = -\left[me^{10}E_0^2(1 + \xi^2)\right]/[8(|m_0| + ½)^3n^3\hbar^6\omega^4], \quad l = |m_0|. \tag{3.48}$$

Like formula (3.42), this expression is valid for $n,l \gg 1$. Again we can see that the energy levels are split and the degeneracy is partially removed. Terms where the magnitude of m_0 is the same, but where the sign differs, will remain degenerate.

Now we proceed to the case of a hydrogen Rydberg atom in a *linearly-polarized* high-frequency laser field $\mathbf{E}(t) = \mathbf{E}_0\cos\,\omega t$. For this case, by using the expression (3.9) for the effective potential as the starting point, Nadezhdin and Oks [4] obtained:

$$U_{\text{eff}}(\text{r}, \theta) = -e^2/r - \left(\gamma_0/r^3\right)(3\cos^2\theta - 1), \quad \gamma_0 = e^4E_0^2/\left(4m_e^2\omega^4\right), \tag{3.49}$$

where θ is the polar angle, that is, the angle between the radius-vector $\mathbf{r}$ of the electron and the z-axis chosen along the vector-amplitude $\mathbf{E}_0$ of the laser field.

It is remarkable that the effective potential from equation (3.49) is mathematically equivalent to the effective potential of a satellite moving around an oblate planet, such as, e.g., the Earth. Indeed, because of this slightly flattened shape, the potential energy $U(r)$ of a satellite in the gravitational field of the Earth slightly differs from the potential energy $U_0(r) = GmM/r$ it would have if the Earth were a sphere. The perturbed potential energy $U(r)$ can be *approximately* represented in the form:

$$U(r, \phi) = -(GmM/r)[1 - I_2(R/r)^2P_2(\sin\phi)]. \tag{3.50}$$

(Formula (3.50) is approximate because it represents only the first two terms of the expansion of the potential energy in inverse powers of r.) Here $G = 6.6726 \times 10^{-8}$ cm^3 (s^{-2} g^{-1}) is the gravitational constant (one of the fundamental constants of Nature); m and M are the masses of the satellite and the Earth, respectively; $I_2 = -1.082 \times 10^{-3}$ is a constant related to the slight difference between the equatorial and polar diameters of the Earth; $R = 6.378 \times 10^8$ cm is the equatorial radius of the Earth; θ is the geographical latitude of the satellite at any point of its orbit; $P_2(\sin\phi)$ is one of the Legendre polynomials:

$$P_2(\sin\phi) = (3\sin^2\phi - 1)/2. \tag{3.51}$$

For the potential energy from equation (3.50) there exists an exact analytical solution for the satellite motion. Details can be found, for example, in section 1.7 of Beletsky's book [8]. The following outcome is similar, but not identical to the case discussed previously in this chapter, where a satellite moves around a prolate planet.

The elliptical orbit of the satellite undergoes two types of the precession simultaneously, but *without changing its shape*. The first one is the precession of the orbit in its plane—pretty much like that depicted in figure 3.2. It occurs with the angular frequency

$$\Omega_{\text{precession in plane}} = (3I_2/4)(R/p)^2(1 - 5\cos^2 i), \tag{3.52}$$

where i is the *inclination*, that is, the angle between the plane of the satellite orbit and the equatorial plane of the Earth. The quantity p in formula (3.52) is related to parameters of the unperturbed elliptical orbit of the satellite, as follows

$$p = 2r_{\min}r_{\max}/(r_{\min} + r_{\max}), \tag{3.53}$$

where $r_{\min}$ and $r_{\max}$ are, respectively, the minimum and maximum distances of the satellite from the center of the Earth.

Interestingly enough, for $i = 63.4°$, one gets $1 - 5\cos^2 i = 0$, so that $\Omega_{\text{precession in plane}}$ vanishes. This means that satellites lunched at this inclination do not undergo the precession in the plane of the orbit.

The second simultaneous precession is the precession of the plane of the satellite orbit, similar to what was illustrated in figure 3.1. The precession frequency is

$$\Omega_{\text{precession of plane}} = (3I_2/2)(R/p)^2 \cos i. \tag{3.54}$$

At $i = 90°$, one gets $\Omega_{\text{precession of plane}} = 0$. For such inclination, which corresponds to the plane of orbit perpendicular to the Earth equator, there is no precession of the plane of orbit. These are so-called polar satellites.

Thus, the fact that the Earth is slightly flattened, does not affect the elliptical shape of the satellite orbit or the inclination of the orbit. It does result, generally, in two simultaneous precessions of the orbit with frequencies given by formulas (3.52) and (3.54).

Taking into account that the geographical latitude ϕ in formula (3.50) and the polar angle θ in formula (3.49) are related as $\phi = \pi/2 - \theta$ (where $\pi/2$ is the 90° angle expressed in radians), so that $\sin^2 \phi = \cos^2 \theta$, and that the quantity I_2 in formula (3.50) is negative (so that $I_2 = -|I_2|$), formula (3.50) can be re-written in the form

$$U(r, \theta) = -(\text{GmM}/r) - (\text{GmM}|I_2|R^2/r^3)(3\cos^2\theta - 1)/2. \tag{3.55}$$

The comparison of formulae (3.49) and (3.55) shows that the potential energy of the electron in a hydrogen Rydberg atom under a high-frequency linearly-polarized laser field is in fact mathematically equivalent to the potential energy of a satellite around the oblate Earth. Indeed, if in formula (3.55) one were to substitute GmM by e^2 and $|I_2|R^2$ by $2\gamma/e^2$, one would obtain formula (3.49).

Therefore, the motion of the electron in a hydrogen Rydberg atom under a high-frequency linearly-polarized laser field can be described in the same way as the motion of a satellite around the oblate Earth. Namely, first of all, the shape of the elliptical orbit and the angle between the orbital plane and the plane perpendicular to the laser field amplitude $\mathbf{E}_0$ would not be affected by the laser field: they would remain the same in the course of time. The elliptical orbit of the electron undergoes two precessions simultaneously: the elliptical orbit precesses in its own plane with the frequency $\Omega_{\text{precession in plane}}$ (similarly to the depiction in figure 3.2) and the plane of the orbit precesses around the laser field amplitude $\mathbf{E}_0$ with the frequency $\Omega_{\text{precession of plane}}$ (similarly to the depiction in figure 3.1), both frequencies being proportional to E_0^2.

There is an interesting fact about the conserved quantities for both of these physical systems (just like in the case of the circular polarization of the laser field, discussed previously in this chapter). If the motion of a particle is characterized by a potential energy that depends only on the magnitude r of the radius-vector $\mathbf{r}$ of the particle, but not on the direction of vector $\mathbf{r}$, then all directions in space are equivalent for such a system: the system is said to have the *spherical symmetry*. As the consequence of the spherical symmetry, the angular momentum vector $\mathbf{M}$ is conserved both by the magnitude and by the direction for such a system. If a linearly-polarized laser field is applied to such an atomic system, then there is no longer the equivalence of all directions in space: the potential energy now depends not only on the magnitude r of the radius-vector $\mathbf{r}$, but also on the angle θ between the vector $\mathbf{r}$ and the laser field amplitude vector $\mathbf{E}_0$. The latter vector defines the preferred direction in space. Only all directions in the plane perpendicular to the vector $\mathbf{E}_0$ remain equivalent: the symmetry of the system is said to be reduced from spherical to axial. As the consequence, only the angular momentum projection M_z on the vector $\mathbf{E}_0$ remains conserved.

Similarly, because the Earth is not a sphere, but the oblate spheroid, the axis connecting the two poles defines the preferred direction in space. Only the angular momentum projection M_z on this direction remains conserved.

However, for the specific form of the potential energy dependence on r and θ, as in equations (3.49) and (3.55), in addition to the exact conservation the angular momentum projection M_z, there is also the *conservation of the square of the angular momentum* M^2. The latter quantity is proportional to the area of the orbit. Therefore, the fact that the shape of the elliptical orbit is not affected by the perturbation (by the laser field for the atomic system of by the spheroidness of the Earth in the case of the satellite) signifies the conservation of M^2.

For completeness we note another physical system equivalent to the 'satellite—spheroidal planet' system. Namely, nuclei of heavy atoms can have either the shape of a prolate spheroid or the shape of the oblate spheroid. So, the motion of the electron in a hydrogen Rydberg atom under a high-frequency laser field of the linear or circular polarizations has also the analogy with the motion of the electron in heavy hydrogenic ions.

It should be emphasized that there are hydrogenic atoms and ions where the negatively charged particle orbiting the nucleus is not the electron, but the muon. Muonic atoms and ions are more sensitive to the nuclear shape than the electronic atoms and ions. In particular, in paper [9] it was shown how the shift of spectral lines of muonic hydrogenic ions can serve for the experimental determination of the parameters of the nuclear shape (such as, for example, the parameter analogous to the constant I_2 in equation (3.50)).

The bottom line is that all the three above physical systems exhibit a *higher than geometrical symmetry*. This is an important result in its own right.

It should be emphasized that for the type of the potential energies given by equations (3.49) or (3.55), the square of the angular momentum M^2 is conserved exactly. However, equations (3.49) and (3.55) were obtained by neglecting some corrections (much smaller than the second term in (3.49) or in (3.55)). Since equations (3.49) and (3.55) are approximate, so is the conservation of M^2. Nevertheless, any additional quantity that is conserved (whether exactly or approximately), is physically important.

Let us discuss now the advantages of this analytical method. We considered a hydrogen atom in a high frequency field

$$\omega \gg \omega_{kn}, \tag{3.56}$$

where ω_{kn} is an atomic transition frequency. In the standard time-dependent perturbation theory, the transition probability can be written as (see, e.g., textbook [10])

$$(\mathrm{d}E_0)^2/(\hbar^2\omega^2) \equiv \omega_E{}^2/\omega^2, \tag{3.57}$$

where $\omega_E \equiv (\mathrm{d}E_0)/\hbar$ may be interpreted as the peak frequency of precession of the dipole moment of the system. Since the atomic dipole moment can be estimated as $d \sim n^2\hbar^2/[me^2]$, then

$$\omega_E \sim n^2\hbar E_0/(me). \tag{3.58}$$

Thus we arrive at the small parameter employed by standard perturbation theory

$$\varepsilon_{\mathrm{pt}} = \omega_E{}^2/\omega^2 < 1. \tag{3.59}$$

In the advanced method of separating rapid and slow subsystems, the small parameter can be found from the ratio of the energy correction to the unperturbed energy

$$\varepsilon_{\mathrm{our}} = \Delta E/E^{(0)}, \; E^{(0)} = me^4/(2\hbar^2n^2). \tag{3.60}$$

Using the above results for ΔE, we find

$$\Delta E/E^{(0)} \sim e^6E_0{}^2/(\hbar^4n^4\omega^4) = \varepsilon_{\mathrm{our}} \tag{3.61}$$

or in atomic units

$$\varepsilon_{\mathrm{our}} = E_0{}^2/(n^4\omega^4) \sim [\omega_E/(n^4\omega^2)]^2 \sim \varepsilon_{\mathrm{pt}}{}^2[\Omega/(n\omega)]^2, \tag{3.62}$$

where $\Omega = (E_{n+1}{}^{(0)} - E_n{}^{(0)})/\hbar = n^{-3}$ is the Kepler frequency. Thus, for the high frequency case ($\Omega \ll \omega$), even when the method of the standard perturbation theory is no longer applicable, i.e. when $\varepsilon_{\mathrm{pt}} \sim 1$, the advanced method of separating rapid and slow subsystems remains valid.

Finally we discuss the connection between the above results and the transition to chaos. Multidimensional problems, like those presented above, play the crucial role in the fascinating transition regime between classical and quantum mechanics. While experiments [11, 12] stimulated this interest, it is also true that the usual tools of analysis which helped the impressive work in the ionization of hydrogen with linearly-polarized radiation are of no use in multiple dimensions and other ways of investigating such problems must be found. In this context, we addressed the dynamics of Rydberg electrons placed in high-frequency circularly-polarized and elliptically-polarized microwave fields in this chapter. We mapped these problems onto problems well-known from celestial mechanics, discovering approximate constants of the motion in the process. We showed how the dynamics of a Rydberg electron whose orbital plane differs from the plane of polarization of circularly-polarized radiation can be mapped on the problem of a satellite orbiting a prolate planet. Although the angular momentum precesses, its magnitude is an approximate constant. Similarly, when the electron's orbit plane coincides with the plane of elliptically-polarized microwaves, it moves like a satellite orbiting a slightly non-spherical planet in its equatorial plane. In this case, the z-component of the angular momentum emerges as a hidden symmetry.

The high-frequency case has received less attention than the low-frequency, resonant cases, and therefore it is desirable to explore the implications of our results for chaos and ionization in these systems. We identified the precession frequency as one of the important parameters of the dynamics, and we expect the onset of chaos when the precession frequency f_e is of the order of the Kepler frequency $1/n^3$.

In the case of circular polarization, we see from equations (3.39), (3.40), (3.42) that the ratio of the precession frequency f_e to the Kepler frequency is

$$f_e n^3 \sim E_0^2/(\omega^4 l^4), \tag{3.63}$$

where l is the angular momentum. When this ratio is of the order of 1, a quasicontinuum forms in the energy spectrum, and the electron diffuses through it into the continuum, leading to ionization. This takes place approximately when

$$E_0 \sim \omega^2 l^2 \equiv E_{\mathrm{nt}}, \tag{3.64}$$

where 'nt' stands for the 'nonresonant threshold'. In contrast, the best-known chaos thresholds are obtained using the resonance overlap criterion of Chirikov [13] (a review covering frequency ranges, polarizations and various initial orbits can be found, e.g., in paper [14]). After giving the critical peak electric field strength E_{rt} at which to expect large-scale stochasticity in the case of linear polarization for orbits of arbitrary shape, these authors also presented the results of calculations for such orbits in the plane of polarization of the circular polarization field. They found that for the overlap caused by the main resonance (corresponding to the ratio between driving and Kepler frequencies, k, being equal to unity), or for the high-frequency linear polarization or circular polarization cases, the thresholds obtained in paper [14] are all of the form

$$E_{\mathrm{rt}} \sim 1/n^4, \tag{3.65}$$

where 'rt' stands for the 'resonant threshold'—the individual cases are distinguished by the proportionality constant, which, in the case of high-frequency circular polarization, is a slowly varying function of the initial angular momentum l. Their general conclusions were that the critical field decreases with increasing angular momentum—more rapidly for linear polarization than for circular polarization. There was good agreement between the critical field of 0.01 (in scaled units) numerically obtained in paper [15], which approaches the earlier value from paper

[16], and the critical field of 0.0109 obtained by a so-called 'renormalization' approach in paper [17], seemingly without being aware of the review by article [14]. Returning to the findings presented in this chapter, it is obviously possible to have

$$E_{\rm nt} \ll E_{\rm rt}, \tag{3.66}$$

if

$$l \ll 1/(\omega n^2), \tag{3.67}$$

In the notation of review [14]

$$\omega = k/n^3, \ k \gg 1, \tag{3.68}$$

we can re-write equation (3.67) as

$$l \ll n/k = l_{\rm nt}. \tag{3.69}$$

Since $k \gg 1$ in our case, the nonresonant threshold value of the angular momentum defined in equation (3.69) is much smaller than n:

$$l_{\rm nt} \ll n. \tag{3.70}$$

Thus we conclude that for sufficiently low l, the nonresonant threshold value of the microwave amplitude can be much smaller than the corresponding resonant threshold value.

This conclusion does not replace the results from review [14], but rather it complements them as follows. On page 561 of review [14] it was written that orbits exhibit stochastic behavior only for

$$l < l_{\rm rt} = (3/\omega)^{1/3} \sim n/k^{1/3}. \tag{3.71}$$

It is easy to see that

$$l_{\rm nt}/l_{\rm rt} \sim 1/k^{2/3} \ll 1. \tag{3.72}$$

Thus, the following picture emerges. For l such that

$$l_{\rm nt} < l < l_{\rm rt}, \tag{3.73}$$

the resonant terms dominate in the ionization and determine the threshold value of E_0. However, for l such that

$$l \ll l_{\rm nt}, \tag{3.74}$$

the nonresonant terms dominate in the ionization, resulting in a lower threshold value of E_0, compared to the resonance approximation.

The connection between the chaos threshold and the initial orbit parameters in elliptical polarization is, predictably, more complicated. In the planar case of elliptical polarization, from equations (3.47), (3.48) the ratio between the precession and Kepler frequencies is the following

$$f_e n^3 \sim E_0^2(1 + \mu^2)/(\omega^4 l^4) = \varepsilon_0^2(1 + \mu^2)^2/(\omega^4 l^4), \tag{3.75}$$

where ε_0 is the microwave peak field defined in equation (3.13).

In both our cases, the threshold E_{crit} is proportional to l^2, so that under the high frequency microwave, low angular momentum, high-eccentricity orbits have the lowest threshold. Of course, the validity of the quasiclassical approach requires that $l \gg 1$, therefore, in the high frequency case, the orbits with the lowest threshold have l in the range of

$$n \gg l \gg 1, \tag{3.76}$$

rather than $l \sim n/2$ as in the resonant cases.

Secondly, from the above formula (3.64) it is seen that in the high frequency case, at a fixed value of l, it is easier to ionize with circularly polarized microwaves ($\mu = 1$) than with the linearly polarized microwaves ($\mu = 0$)—in distinction to the lower frequency case. In paper [17] the resonance overlap criterion was applied to a reduced-dimensionality version of elliptical polarization in the plane of polarization ((x,y) in their case) in which the initial angular momentum and the angle φ between the x-axis and the Runge–Lenz vectors are constant to first order in the field. It was found in paper [17] that the threshold for ionization is due to the overlap of neighboring substates (m, l) and (m + 1, 1) and they concluded that orbits with eccentricity ~0.6 are the first to ionize—indeed, for all the ellipticity degrees, orbits with medium eccentricity have the lowest threshold for unbounded diffusion, as averaging their results over the angle φ shows: the scaled ionization threshold dips as low as 0.015 for eccentricity ~0.6, rising to 0.045 for lower values of eccentricity.

Thus, there are remarkable distinctions between resonant and non-resonant cases revealed in this chapter.

References

[1] Kapitza P L 1951 *Sov. Phys. JETP* **21** 588

[2] Kapitza P L 1951 *Uspekhi Fiz. Nauk* **44** 7

[3] Landau L D and Lifshitz E M 1960 *Mechanics* (Oxford: Pergamon) section 30

[4] Nadezhdin B B and Oks E 1986 *Sov. Tech. Phys. Lett.* **12** 512

[5] Oks E, Davis J and Uzer 2000 *J. Phys. B: At. Mol. Opt. Phys* **33** 207

[6] Roy A E 1978 *Orbital Motion* (Bristol: Adam Hilger)

[7] Goldstein H 1980 *Classical Mechanics* (Reading, MA: Addison-Wesley)

[8] Beletsky V V 2001 *Essays on the Motion of Celestial Bodies* (Basel: Birkhäuser/Springer)

[9] Oks E 2018 *Atoms* **6** 14

[10] Landau L D and Lifshitz E M 1965 *Quantum Mechanics* (Oxford: Pergamon)

[11] Bellerman M R W, Koch P M, Mariani D R and Richards D 1996 *Phys. Rev. Lett.* **76** 892

[12] Bellerman M R W, Koch P M and Richards D 1997 *Phys. Rev. Lett.* **78** 3840

[13] Chirikov B V 1979 *Phys. Rep.* **52** 265

[14] Delone N B, Krainov V P and Shepelyansky D I 1983 *Sov Phys.—Uspekhi* **26** 551

[15] Howard J E 1992 *Phys. Rev.* A **46** 364

[16] Mostowski J and Sanchez-Mondragon J J 1979 *Opt. Commun.* **29** 293

[17] Sasha K and Zakrzewski J 1997 *Phys. Rev.* A **55** 568

Chapter 4

Rydberg states of muonic-electronic helium atoms or helium-like ions

Systems where one of the electrons is substituted by the heavier lepton μ^-, i.e., muonic atoms and molecules, are subjects of studies having several applications. The first application is the fusion catalyzed by muon (see, e.g., [1–3] and references therein). This is because a muon replacing the electron either in the *dde*-molecule (D_2^+) or in the *dte*-molecule, leads to the decrease of the equilibrium internuclear distance by a factor of about 200. At so dramatically decreased internuclear distances, the fusion can occur with a significant probability. This has been observed in $dd\mu$ or even with a higher rate in $dt\mu$ [1–3].

The second application has to do with a laser-control of nuclear processes. The corresponding studies were performed in the context of the interaction of muonic molecules with superintense laser fields [4].

The third application relates to the search for strongly interacting massive particles (SIMPs). SIMPs were suggested as candidates for dark matter and as candidates for the lightest supersymmetric particle (see, e.g., [5] and references therein). SIMPs could become bound to the atomic nuclei, leading to anomalously heavy isotopes of known elements. Since SIMPs drastically increase the nuclear mass, their presence in the nucleus practically eliminates the well-known reduced mass correction in a hydrogenic atom. It is easier to observe this effect in muonic atoms (rather than in electronic atoms) because the muon's much larger mass (compared to the electron) significantly enhances the reduced mass correction [5]. Such an effect may be observable in astrophysical objects [5].

In the present paper we consider helium atoms or helium-like ions (though the primary focus is on helium atoms) where one of the electrons is substituted by a muon and where both leptons are in Rydberg states. In particular, the muon is

doi:10.1088/2053-2563/ab3db0ch4

assumed to be in a circular Rydberg state[1]. However, no assumption is made concerning the shape of the electron orbit or its orientation with respect to the plane of the muon orbit.

In the muonic-electronic helium atom or helium-like ion, the muon motion is characterized by a much higher frequency than the electron motion. Therefore, we apply the analytical method based on separating rapid and slow subsystems. We demonstrate that the electron moves in an effective potential that is mathematically equivalent to the potential, in which a satellite moves around an oblate planet (such as, e.g., the oblate Earth).

Based on this celestial analogy, we show that the 'unperturbed' elliptical orbit of the electron undergoes simultaneously two precessions. One of them is the precession of the electron orbit in the plane of the orbit, another is the precession of the orbital plane of the electron around the axis of symmetry of the muonic orbit. These two precessions occur with different frequencies and we provide analytical expressions for these two precession frequencies. We emphasize that the shape of the elliptical orbit of the Rydberg electron does not change in the course of these two precessions. This means that the square of the angular momentum of the Rydberg electron is conserved (approximately), which is the manifestation of the *hidden symmetry* of the system.

So, we consider a system consisting of a muon, an electron, and a nucleus of charge Z. Both the muon and the electron are in Rydberg states, so that their principal quantum numbers (n_μ for the muon and n_e for the electron) are much greater than unity. Because of the large difference between the muon mass m_μ and the electron mass m_e ($m_\mu/me = 206.8$), the muon is much closer to the nucleus than the electron. We assume that the muon is in a circular Rydberg state, so that its angular momentum quantum number has the maximum possible value of $(n_\mu - 1)$.

The frequency of the revolution of the muon and of the nucleus around their center of mass is

$$\Omega = m_{\mu r} Z^2 e^4 / \left(n_\mu{}^3 \hbar^3\right). \tag{4.1}$$

Here,

$$m_{\mu r} = m_\mu M_{\text{nucl}} / (m_\mu M_{\text{nucl}}) \tag{4.2}$$

is the reduced mass of the pair 'nucleus-muon', M_{nucl} being the nuclear mass. In particular, for helium atoms ($Z = 2$), one has $M_{\text{nucl}}/m_\mu = 35.28$, so that in atomic units ($\hbar = m_e = e = 1$) the reduced mass $m_{\mu r} = 201.1$.

In the first approximation, the Rydberg electron moves in the Coulomb field of the effective charge $(Z - 1)$, so that the Kepler frequency of the Rydberg electron is

[1] We note that circular Rydberg states (CRS) of atoms were studied to a large extent [6–9] theoretically and experimentally for several reasons. First, it was because CRS have long radiative lifetimes and strongly anisotropic collision cross sections, so that they facilitate experiments on inhibited spontaneous emission and cold Rydberg gases [10, 11]. Second, classical CRS are counterparts to fundamentally important quantal coherent states. Third, in the quantal method employing the $1/n$-expansion (n being the principal quantum number), the primary term corresponds to the classical description of (see, e.g. paper [12] and references therein).

$$\omega_K = m_{er}(Z - 1)^2 e^4 / \left(n_e{}^3 \hbar^3\right). \tag{4.3}$$

Here,

$$m_{er} = m_e(m_\mu + M_{\text{nucl}})/(m_e + m_\mu + M_{\text{nucl}}) \tag{4.4}$$

is the reduced mass of the electron in this system. In particular, for helium atoms ($Z = 2$), one has $m_{er} = 0.9999$ in atomic units, so that at any $Z > 1$, for almost all practical purposes it is possible to use $m_{er} = 1$.

From equations (4.1) and (4.3), it is easy to see that the ratio of the frequencies

$$\Omega/\omega_K = m_{\mu r} Z^2 n_e{}^3 / [m_e(Z - 1)^2 n_\mu{}^3] \gg 1 \tag{4.5}$$

if

$$n_\mu / n_e \ll (m_{\mu r} Z^2)^{1/3} / [m_e(Z - 1)^2]^{1/3}. \tag{4.6}$$

In particular, for helium atoms equation (4.6) becomes $n_\mu/n_e \ll 9$ and equation (4.5) becomes

$$\Omega/\omega_K = 804.4\, (n_\mu / n_e)^3. \tag{4.7}$$

For $Z \gg 1$, equation (4.6) becomes $n_\mu/n_e \ll 6$ and equation (4.5) becomes

$$\Omega/\omega_K = 206.8\, (n_\mu / n_e)^3. \tag{4.8}$$

Thus, it is seen that, e.g., at $n_\mu/n_e \sim 1$, the frequency of the revolution of the muon exceeds the Kepler frequency of the electron by two or three orders of magnitude (obviously, the ratio Ω/ω_K is even greater for $n_\mu/n_e < 1$). So, indeed the pair 'nucleus-muon' constitutes a rapid subsystem, while the electron is a slow subsystem.

Therefore, for finding the electron orbit in the second approximation, one should perform the averaging over the rapid subsystem. Then this brings up the following physical picture. The electron perceives both the muon and the nucleus as circular rings of the radii R_μ and R_{nucl}, respectively, the muon charge being uniformly distributed over the ring of the radius R_μ and the nuclear charge being uniformly distributed over the ring of the radius R_{nucl}. The ratio of the radii is

$$R_\mu / R_{\text{nucl}} = M_{\text{nucl}} / m_\mu \gg 1. \tag{4.9}$$

In particular, for helium atoms one has $R_\mu/R_{\text{nucl}} = 35.28$.

So, the effective potential energy for the electron consists of two terms $U^{(1)}$ and $U^{(2)}$. The primary term $U^{(1)}$ is the effective Coulomb interaction with the nucleus and the muon 'combined in one':

$$U^{(1)} = -(Z - 1)e^2/r. \tag{4.10}$$

The second term $U^{(2)}$ is due to the quadrupole interaction (since the dipole moment of the muonic and nuclear 'rings' is zero). In the spherical polar coordinates with the z-axis being the axis of the symmetry of the muonic 'ring', its contribution $U^{(2)}{}_\mu$ to the quadrupole potential energy is

$$U^{(2)}{}_{\mu} = -e^2 R_{\mu}{}^2 (3\cos^2\theta - 1)/(4r^3). \tag{4.11}$$

Here r is the absolute value of the radius-vector of the electron and θ is the polar angle of the electron.

The corresponding quadrupole contribution of the nuclear 'ring' is

$$U^{(2)}{}_{\text{nucl}} = Ze^2 R_{\text{nucl}}{}^2 (3\cos^2\theta - 1)/(4r^3). \tag{4.12}$$

Taking into account equation (4.9), the ratio of absolute values $|U^{(2)}{}_{\mu}/U^{(2)}{}_{\text{nucl}}|$ can be expressed as follows:

$$\left|U^{(2)}{}_{\mu}/U^{(2)}{}_{\text{nucl}}\right| = R_{\mu}{}^2/\left(Z\,R_{\text{nucl}}{}^2\right) = M_{\text{nucl}}{}^2/\left(Zm_{\mu}{}^2\right) \gg 1. \tag{4.13}$$

In particular, for helium atoms $|U^{(2)}{}_{\mu}/U^{(2)}{}_{\text{nucl}}| = 622 \gg 1$. For $Z > 2$, the ratio $|U^{(2)}{}_{\mu}/U^{(2)}{}_{\text{nucl}}|$ is even greater than for $Z = 2$: since M_{nucl} scales roughly linearly with Z, then $|U^{(2)}{}_{\mu}/U^{(2)}{}_{\text{nucl}}|$ also scales roughly linearly with Z. Therefore, for almost all practical purposes the contribution of the nuclear 'ring' to the quadrupole potential energy can be disregarded.

Thus, the effective potential energy of the Rydberg electron can be represented in the form

$$U_{\text{eff}} = -(Z - 1)e^2/r - e^2 R_{\mu}{}^2 (3\cos^2\theta - 1)/(4r^3). \tag{4.14}$$

Let us compare U_{eff} from equation (4.14) with the following potential energy U_{E} for a satellite around the oblate Earth (see, e.g., Beletsky book [13])

$$U_{\text{E}} = -GmM_{\text{E}}/r - GmM_{\text{E}}\,|I_2|\,R^2(3\cos^2\theta - 1)/(2r^3). \tag{4.15}$$

Here, m is the mass of the satellite, M_{E} is the Earth mass, R is the equatorial radius of the Earth, and I_2 is a constant characterizing the relative difference between the equatorial and polar diameters of the Earth.

If in equation (4.15) we were to redefine (i.e., bring into the correspondence)

$$GmM_{\text{E}} = (Z - 1)e^2, \quad |I_2|\,R^2 = e^2 R_{\mu}{}^2/(2GmM_{\text{E}}) = R_{\mu}{}^2/[2(Z - 1)], \tag{4.16}$$

then the right side of equation (4.15) would become identical to the right side of equation (4.14). So, the problem of the motion of the Rydberg electron in muonic-electronic helium atoms or helium-like ions is indeed mathematically equivalent to the problem of the motion of a satellite around an oblate planet (such as, e.g., the oblate Earth). The solution for the latter problem is well-known (see, e.g., Beletsky's book [13]). Namely, the elliptical orbit of the satellite undergoes two types of the precession simultaneously, but *without changing its shape*. The first one is the precession of the orbit in its plane with the angular frequency

$$\omega_{\text{E,precession in plane}} = (3I_2/4)(R/p)^2(1 - 5\cos^2 i)\omega_{\text{E,K}}, \tag{4.17}$$

where

$$\omega_{E,K} = \left[G(M_E + m_s)/A_s^3\right]^{1/2} \quad (4.18)$$

is the Kepler frequency of the satellite, m_s and A_s being the satellite mass and the major semi-axis of its unperturbed elliptical orbit, respectively. In equation (4.17), i is the *inclination*, that is, the angle between the plane of the satellite orbit and the equatorial plane of the Earth. The quantity p (semi-latus rectum) in formula (17) is related to parameters of the unperturbed elliptical orbit of the satellite as follows

$$p = 2r_{\min} r_{\max} / (r_{\min} r_{\max}), \quad (4.19)$$

where $r_{\min}$ and $r_{\max}$ are, respectively, the minimum and maximum distances of the satellite from the center of the Earth.

The second simultaneous precession is the precession of the plane of the satellite orbit around the axis along the polar diameter of the Earth. The frequency of this precession is

$$\omega_{E,\text{precession of plane}} = (3I_2/2)(R/p)^2(\cos i)\omega_{E,K}. \quad (4.20)$$

Thus, with the help of the correspondence formulas from equation (4.16) we can immediately describe the motion of the Rydberg electron in muonic-electronic helium atoms or helium-like ions as follows. While doing this, we take into account the relation between the semi-latus rectum p of the unperturbed elliptical orbit of the Rydberg electron and its angular momentum M

$$p = M^2/[(Z - 1)m_e e^2] = (l_e + 1/2)^2\hbar^2/[(Z - 1)m_e e^2], \quad (4.21)$$

where l_e is the angular momentum quantum number of the electron.

So, the elliptical orbit of the electron undergoes two types of the precession simultaneously, but *without changing its shape*. The first one is the precession of the orbit in its plane with the angular frequency

$$\omega_{\text{precession in plane}} = (3/8)[(Z - 1)/Z^2](1 - 5\cos^2 i)[n_\mu/(l_e + 1/2)]^4(m_{er}/m_{\mu r})^2\omega_K, \quad (4.22)$$

where the Kepler frequency ω_K of the Rydberg electron is given by equation (4.3). For $i = 63.4°$, one gets $1 - 5\cos^2 i = 0$, so that $\omega_{\text{precession in plane}}$ vanishes. This means that if the unperturbed orbital plane of the Rydberg electron has this inclination, then there is no precession within the plane of the orbit.

The second simultaneous precession is the precession of the orbital plane of the Rydberg electron around the axis of the symmetry of the muonic 'ring'. The frequency of this precession is

$$\omega_{\text{precession of plane}} = (3/4)[(Z - 1)/Z^2](\cos^2 i)[n_\mu/(l_e + 1/2)]^4(m_{er}/m_{\mu r})^2\omega_K. \quad (4.23)$$

At $i = 90°$, one gets $\omega_{\text{precession of plane}} = 0$. For such inclination, which corresponds to the orbital plane of the Rydberg electron being perpendicular to the orbital plane of the Rydberg muon, there is no precession of the plane of orbit around the axis of the symmetry of the muonic 'ring'.

Interestingly enough, it turns out that the above the problem of the motion of the Rydberg electron in muonic-electronic helium atoms or helium-like ions is mathematically equivalent also to another problem from atomic physics, as follows. In paper [14] the authors considered a hydrogen Rydberg atom in a linearly-polarized electric field of a high-frequency laser radiation $\mathbf{E}(t) = \mathbf{E}_0 \cos \omega_L t$, the laser frequency ω_L being much greater than the Kepler frequency of the Rydberg electron. By applying the analytical method of separating the rapid and slow subsystems, they showed that the perturbed motion of the atomic electron that occurs is characterized by the following effective potential energy (in spherical polar coordinates with the z-axis along vector $\mathbf{E}_0$)

$$U_{\text{eff}} = -e^2/r - (\gamma/r^3)(3\cos^2\theta - 1), \quad \gamma = e^4 E_0^2/\left(4m_e^2\omega^4\right). \tag{4.24}$$

It is seen that the effective potential energy U_{eff} from equation (4.24) is indeed mathematically equivalent to the effective potential energy U_{eff} from equation (4.14) for the motion of the Rydberg electron in muonic-electronic helium atoms or helium-like ions.

The authors of paper [14] also pointed out that while the geometrical symmetry of the above physical systems is axial, so that only the projection M_z of the angular momentum on the axis of the symmetry is exactly conserved, there is also an approximate conservation of the square of the angular momentum M^2.[2] The latter quantity is proportional to the area of the orbit. Therefore, the fact that the shape of the elliptical orbit is not affected by the perturbation is the manifestation of the conservation of M^2. This means that the above physical systems have a *higher than geometrical symmetry* (sometimes called *hidden symmetry*), which is a *counter-intuitive* result. Only a relatively small number of physical systems have a hidden symmetry. So, the fact that a muonic-electronic helium atom or helium-like ion in Rydberg states possesses the hidden symmetry should be of a *general physical interest*.

For completeness we mention that in the corresponding quantum problem of a hydrogen atom in a linearly-polarized high-frequency laser field, the analytical solution for which was presented in paper [15], there is also a manifestation of the hidden symmetry. Namely, despite the states of the unperturbed system being degenerate, it turned out to be possible to apply the simpler formalism of the perturbation theory for non-degenerate states. This was because the non-diagonal elements of the perturbation (the perturbation being the second term in U_{eff} from equation (4.24)) turned out to be zeros.

In summary, we considered muonic-electronic helium atoms or helium-like ions (with the primary focus on helium atoms), where both the muon and the electron are in Rydberg states, the muon being in a circular Rydberg state. We showed that the

[2] It should be noted that for the type of the potential energy given by formulas (4.14) or (4.24), the square of the angular momentum M^2 is conserved exactly. However, formulas (14.4) and (4.24) were obtained by neglecting some corrections (much smaller than the second term in (4.14) or in (4.24)). Since formulas (4.14) and (4.24) are approximate, so is the conservation of M^2. Nevertheless, any additional quantity that is conserved (whether exactly or approximately), is physically important.

subsystem 'nucleus—muon' can be treated as the rapid subsystem, while the electron is the slow subsystem.

We demonstrated that the motion of the Rydberg electron occurs in a modified Coulomb potential, where the second term is due to the quadrupole interaction with muonic 'ring'. We showed that the effective potential energy of the Rydberg electron is mathematically equivalent to the potential energy of a satellite moving around an oblate planet (e.g., the oblate Earth). Based on this, we demonstrated that the unperturbed orbital plane of the Rydberg electron undergoes simultaneously two different precessions: one is the precession within the orbital plane, the other the precession of the orbital plane around the axis of the muonic 'ring'. We provided analytical expressions for the frequencies of both precessions.

We emphasized that the shape of the elliptical orbit of the Rydberg electron is not affected by the perturbation, which is the manifestation of the (approximate) conservation of the square of the angular momentum of the Rydberg electron. This means that the above physical systems have a *higher than geometrical symmetry* (also known as *hidden symmetry*), which is a *counterintuitive* result of a *general physical interest*.

Finally, we noted that the above problem of the motion of the Rydberg electron in muonic-electronic helium atoms or helium-like ions is mathematically equivalent also to another problem from atomic physics: a hydrogen Rydberg atom in a linearly-polarized electric field of a high-frequency laser radiation.

We note that in papers [16, 17] the authors considered Rydberg states of muonic-electronic negative hydrogen ion, the results being presented in appendix B of this book. In another paper [18] the authors analyzed Rydberg states of muonic-electronic helium-like atoms or ions, the results being presented in appendix C of this book. In all the three papers [16–18], the authors used the method of separating rapid and slow subsystems.

While the systems considered in the present chapter and in paper [18] are identical, here is an important distinction between the approaches. The system considered in the present chapter is truly stable, the electron orbit is generally elliptical, but the relatively small influence of the electron on the muon was neglected. In paper [18] the influence of the electron on the muon was taken into account; however, in the rotating frame used in paper [18] the motion of the muon becomes only metastable (not truly stable); besides, only the circular orbits of the electron were considered in paper [18].

References

[1] Ponomarev L I 1990 *Contemp. Phys.* **31** 219

[2] Nagamine K 2001 *Hyperfine Interact.* **138** 5

[3] Nagamine K and Ponomarev L I 2003 *Nucl. Phys.* A **721** C863

[4] Chelkowsky C, Bandrauk A D and Corkum P B 2004 *Laser Phys.* **14** 473

[5] Guffin J, Nixon G, Javorsek D II, Colafrancesco S and Fischbach E 2002 *Phys. Rev.* D **66** 123508

[6] Lee E, Farrelly D and Uzer T 1997 *Opt. Express* **1** 221

[7] Germann T C, Herschbach D R, Dunn M and Watson D K 1995 *Phys. Rev. Lett.* **74** 658

[8] Cheng C H, Lee C Y and Gallagher T F 1994 *Phys. Rev. Lett.* **73** 3078
[9] Chen L, Cheret M, Roussel F and Spiess G 1993 *J. Phys.* B **26** L437
[10] Dutta S K, Feldbaum D, Walz-Flannigan A, Guest J R and Raithel G 2001 *Phys. Rev. Lett.* **86** 3993
[11] Hulet R G, Hilfer E S and Kleppner D 1985 *Phys. Rev. Lett.* **55** 2137
[12] Vainberg V M, Popov V S and Sergeev A V 1990 *Sov. Phys. JETP* **71** 470
[13] Beletsky V V 2001 *Essays on the Motion of Celestial Bodies* (Basel: Birkhäuser/Springer) section 1.7
[14] Nadezhdin B B and Oks E 1986 *Sov. Tech. Phys. Lett.* **12** 512
[15] Gavrilenko V P, Oks E and Radchik A V 1985 *Opt. Spectrosc.* **59** 411
[16] Kryukov N and Oks E 2012 *Inter. Rev. Atom. Molecul. Phys.* **3** 17
[17] Kryukov N and Oks E 2013 *Can. J. Phys.* **91** 715
[18] Kryukov N and Oks E 2014 *Can. J. Phys.* **92** 1405

Chapter 5

A circumbinary planet around a binary star in Einstein's general relativity and in Newton's gravity

Studies of planets around binary stars are especially important because it is estimated that approximately one half of them could potentially support life [1–3]. Most of the published analytical results related to a simplified situation where the planet is confined to the plane, in which the two stars orbit one another, thus reducing the problem to two dimensions [1–9].

As for studies of a three-dimensional motion, we can mention paper [10] and references therein concerning analytical results, and paper [11] and references therein concerning simulations. Also, in paper [12] it was shown analytically the possibility of a relatively stable 'cork screw' trajectory of the planet. The cork screw has a shape of a cone coaxial with the interstellar axis. In paper [13] were specified the ranges of parameters required for such trajectory to have a long-time stability.

In the present chapter, we discuss the motion of a circumbinary planet (that is, for the situation where the stars move along circular orbits) in frames of Newton's gravity and in frames of Einstein's general relativity (limiting ourselves by the first nonvanishing relativistic correction). We focus at various types of the precession of the planetary orbit. For the case of Newton's gravity the results are produced by means of the analytical method of separating rapid and slow subsystems. For the latter case we point out that this physical system has analogies both in celestial and quantum mechanics, and that this physical system possesses higher than geometrical symmetry.

Let us start by considering a motion of a light planet of mass m around a heavy central mass $m_0 \gg m$. The equation of the relative motion can be represented in the form

doi:10.1088/2053-2563/ab3db0ch5

$$(m/2)(\mathrm{d}r/\mathrm{d}\tau)^2 = [E^2/(2mc^2) - mc^2/2] + Gm_0m/r - L^2/(2Mr^2) + G(m_0 + m)L^2/(c^2Mr^3). \tag{5.1}$$

Here, τ is the proper time, E is the energy, G is the gravitational constant, L is the angular momentum, and $M = m_0m/(m_0 + m)$ is the reduced mass. So, the radial motion of the planet occurs in the following relativistic effective potential

$$V_r = -Gm_0m/r + L^2/(2Mr^2) - G(m_0 + m)L^2/(c^2Mr^3). \tag{5.2}$$

By introducing notations

$$h = L^2/(2Gm_0mM), \quad g = Gm_0mM/c^2, \tag{5.3}$$

h being the scaled square of the angular momentum, equation (5.2) can be re-written in the form

$$V_{r,s} = -1/r + h/(2r^2) - hg/r^3, \tag{5.4}$$

where

$$V_{r,s} = V_r/(Gm_0m) \tag{5.5}$$

is the scaled relativistic effective potential.

Figure 5.1 shows a three-dimensional plot of the scaled relativistic effective potential $V_{r,s}$ versus both the radial variable r and the scaled square of the angular momentum h for $g = 0.1$.

Figures 5.2 and 5.3 present plots of the scaled relativistic effective potential $V_{r,s}$ versus the radial variable r at $g = 0.1$ for two values of the scaled square of the angular momentum $h = 0.5$ and $h = 0.1$, respectively.

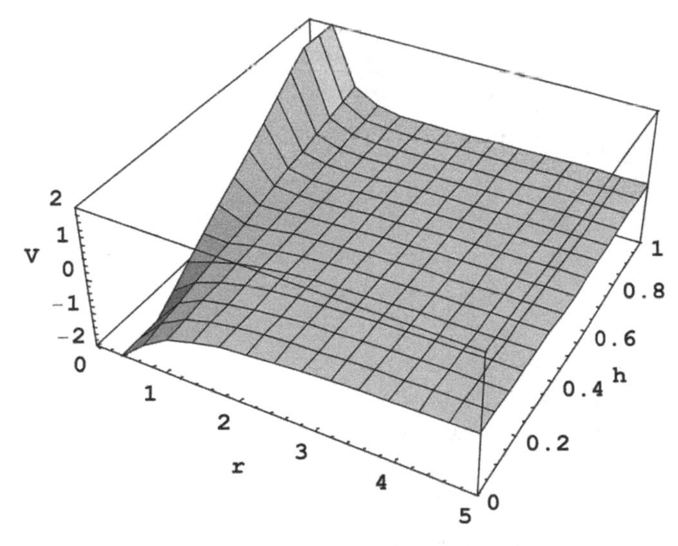

Figure 5.1. Three-dimensional plot of the scaled relativistic effective potential $V_{r,s}$ versus both the radial variable r and the scaled square of the angular momentum h for $g = 0.1$.

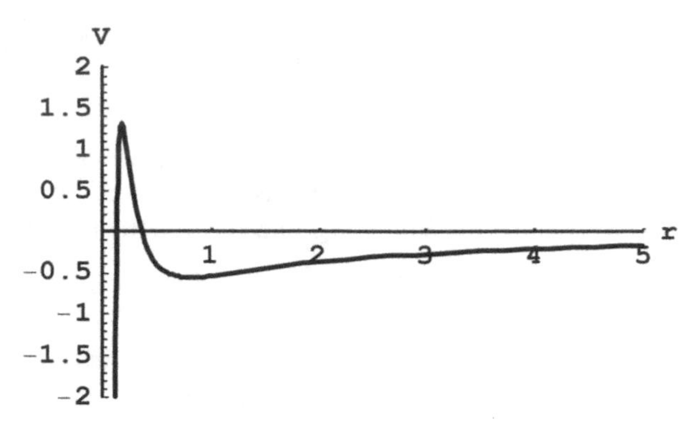

Figure 5.2. The scaled relativistic effective potential $V_{r,s}$ versus the radial variable r at $g = 0.1$ for the scaled square of the angular momentum $h = 0.5$.

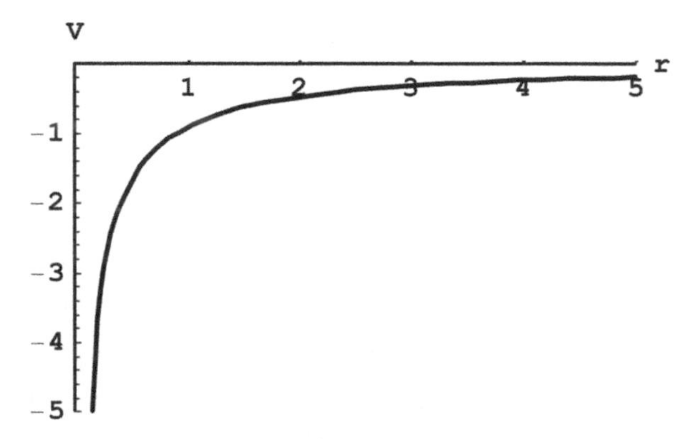

Figure 5.3. The scaled relativistic effective potential $V_{r,s}$ versus the radial variable r at $g = 0.1$ for the scaled square of the angular momentum $h = 0.1$.

From figures 5.1–5.3 it is seen that for the existence of stable planetary orbits, the angular momentum must be sufficiently large. This is the distinction from the non-relativistic case where stable planetary orbits are possible for any non-zero value of the angular momentum.

Due to the third, relativistic term in the effective potential from equation (5.2), the elliptical orbit of the planet undergoes a precession. The relativistic precession angle $\delta\varphi_r$ per one revolution over the elliptical orbit is

$$\delta\varphi_r = 6\pi G(m_0 + m)/[c^2 a(1 - \varepsilon^2)], \tag{5.6}$$

where a and ε are the major semi-axis and the eccentricity of the ellipse, respectively.

Now we proceed to the situation where the heavy central mass is represented by a binary star. We consider the case where the average separation a_0 between the two stars is much smaller than the major semi-axis a of the planetary ellipse:

$$a_0 \ll a. \tag{5.7}$$

Then in the first approximation, the radial motion of the planet can be described by the relativistic effective potential V_r from equation (5.2) after the substitution

$$m_0 = m_1 + m_2, \tag{5.8}$$

where m_1 and m_2 are the masses of the two stars. Then the relativistic precession angle $\delta\varphi$ per one revolution over the elliptical orbit takes the form

$$\delta\varphi_r = 6\pi G(m_1 + m_2 + m)/[c^2 a(1 - \varepsilon^2)]. \tag{5.9}$$

Now let us discuss this problem in frames of Newton's gravity. We limit ourselves by the case of a circumbinary planet, meaning that both stars move in circular orbits around their barycenter, being separated by the fixed distance a_0. The plane of the planetary orbit, generally speaking, is not co-planar with the orbital plane of the stars.

Under the condition (5.7), the Kepler frequency of the stars rotation $\Omega = [G(m_1 + m_2)/a^3]^{1/2}$ is much greater than the Kepler frequency ω of the planet $\omega = [G(m_1 + m_2 + m)/A^3]^{1/2}$. Thus, the binary star can be treated as a rapid subsystem and the circumbinary planet can be treated as a slow subsystem. Therefore, the planet perceives each star as a circular ring of the mass m_k uniformly distributed around the ring (here $k = 1, 2$).

In paper [14] it was shown that the effective potential for the planet in this picture is

$$V_{\text{eff}} = -G(m_1 + m_2)m/r - G\mu \text{ma}^2(3\cos^2\theta - 1)/(4r^3), \tag{5.10}$$

where

$$\mu = m_1 m_2/(m_1 + m_2). \tag{5.11}$$

In equation (5.10), θ is the polar angle counted from the axis of the symmetry of the stellar 'rings'.

The effective potential from equation (5.10) is mathematically equivalent to the potential of a satellite orbiting an oblate planet, such as, for example the Earth. The latter potential can be found, for instance, in book [15]. It is well-known that the elliptical orbit of the satellite engages in two types of precession simultaneously, while the shape of the orbit does not change [15]. First, the satellite orbit precesses in its own plane. The ratio of the Kepler period T of the satellite to the period of this precession t_1 is [15]

$$T/t_1 = (3I_2/4)\{R/[a(1 - \varepsilon^2)]\}^2(1 - 5\cos^2 i), \tag{5.12}$$

where I_2 is a constant characterizing the relative difference between the equatorial and polar diameters of the Earth, R is the equatorial radius of the Earth, and i is the *inclination*, that is, the angle between the plane of the satellite orbit and the equatorial plane of the Earth.

Second, the satellite orbital plane precesses around the axis along the polar diameter of the Earth. The ratio of the Kepler period T of the satellite to the period of that precession t_2 is [15]:

$$T/t_2 = (3I_2/2)\{R/[a(1 - \varepsilon^2)]\}^2 \cos i. \tag{5.13}$$

Similarly, the elliptical orbit of the circumbinary planet engages in two types of precession simultaneously, while the shape of the orbit does not change. First, the satellite orbit precesses in its own plane. The ratio of the Kepler period T of the satellite to the period of this precession t_1 is [14]:

$$T/t_1 = (3/8)\{a_0/[a(1 - \varepsilon^2)]\}^2(1 - 5\cos^2 i)m_1m_2/(m_1 + m_2)^2. \tag{5.14}$$

Second, the satellite orbital plane precesses around the axis of the symmetry of the stellar 'rings'. The ratio of the Kepler period T of the satellite to the period of that precession t_2 is [10, 14]:

$$T/t_2 = (3/4)\{a_0/[a(1 - \varepsilon^2)]\}^2(\cos i)\, m_1m_2/(m_1 + m_2)^2. \tag{5.15}$$

Both of the above physical systems are also mathematically equivalent to a hydrogen Rydberg atom in a linearly-polarized electric field of a high-frequency laser radiation, the latter system being studied analytically in paper [16].

From the geometrical point of view, all the three above physical systems have only the axial symmetry. For systems possessing the axial symmetry, only the projection of the angular momentum on the axis of the symmetry is exactly conserved. However, for all the three above physical systems the square of the angular momentum is also conserved, though approximately. This is manifested by the invariance of the shape of the elliptical orbit during the precessions.

This signifies an *algebraic symmetry* (sometimes called *hidden symmetry*) that is *higher than geometrical symmetry.*

Coming back to the circumbinary planet: so, there is a relativistic precession of the planetary orbit and two non-relativistic precessions of it. How do they compare to each other? First, let us calculate the ratio ω_1/ω_2 of the periods of the two above non-relativistic precessions:

$$\text{rat} = \omega_1/\omega_2 = (1 - 5\cos^2 i)/(2\cos i). \tag{5.16}$$

Figure 5.4 shows this ratio versus the inclination i (radians). It is seen that as the inclination increases, so does this ratio. It changes the sign at $i = 1.11$ radians, which corresponds to 63.4°. The periods/frequencies of the two non-relativistic precessions become equal to each other at $i = 1.28$ radians, which corresponds to 73.1°.

To simplify a further comparison let us consider the particular case where the planetary orbit is in the plane of the star's rotation: $i = 0$. In this case, the frequencies of the two non-relativistic precessions have opposite signs, meaning that they 'work' in opposite directions. The absolute value of the frequency ω_1 is twice as much as the frequency ω_2. So, the effective frequency of the non-relativistic precession is $\omega_{\text{eff}} = \omega_1 + \omega_2$. The ratio of $|\omega_{\text{eff}}|$ to the Kepler frequency ω of the planet is:

$$\omega_{\text{eff}}/\omega = -(3/4)\{a_0/[a(1 - \varepsilon^2)]\}^2 m_1m_2/(m_1 + m_2)^2. \tag{5.17}$$

The angle of the precession $\delta\varphi$, corresponding to the ratio from equation (5.17) is

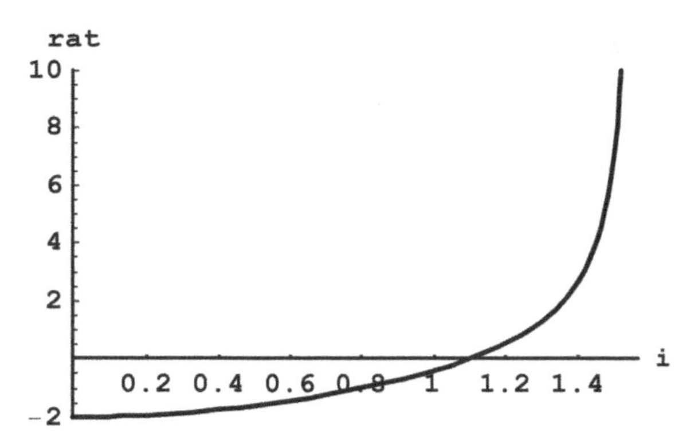

Figure 5.4. The ratio of the periods of the two non-relativistic precessions of the orbital plane of a circumbinary planet versus the inclination i (radians).

$$\delta\varphi = -(3\pi/2)\{a_0/[a(1 - \varepsilon^2)]\}^2 m_1 m_2/(m_1 + m_2)^2. \tag{5.18}$$

Now let us calculate the ratio of the angle of the relativistic precession $\delta\varphi_r$ from equation (5.9) to the angle of the non-relativistic precession $\delta\varphi$ from equation (5.18):

$$\delta\varphi_r/\delta\varphi = -[2G(m_1 + m_2)^2(m_1 + m_2 + m)/(c^2 m_1 m_2 a_0)][a(1 - \varepsilon^2)/a_0]. \tag{5.19}$$

In equation (5.19), the factor in the first bracket is much smaller than unity, while the factor in the second bracket is much greater than unity. Therefore, depending on the interplay of these two factors, the precession of the planetary orbit could be controlled either by the relativistic precession or by the non-relativistic precession.

References

[1] Quintana E V and Lissauer J J *Planets in Binary Star Systems* ed N Naghighipour (Dordrecht: Springer), ch 10 p 265

[2] Fatuzzo M, Adams F C, Gauvin R and Proszkow E M 2006 *Publ. Astron. Soc. Pac.* **118** 1510

[3] David E, Quintana E V, Fatuzzo M and Adams F C 2003 *Publ. Astron. Soc. Pac.* **115** 825

[4] Kaib N A, Raymond S N and Duncan M 2013 *Nature* **493** 381

[5] Desidera S and Barbieri M 2007 *Astron. Astrophys.* **462** 345

[6] Quintana E V, Adams F C, Lissauer J J and Chambers J E 2007 *Astrophys. J.* **660** 807

[7] Turrini D, Barbieri M, Marzari F, Thebault P and Tricarico P 2005 *Mem. S. A. It. Suppl.* **6** 172

[8] Holman M J and Wiegert P A 1999 *Astron. J.* **117** 621

[9] Innanen K A, Zheng J Q, Mikkola S and Valtonen M J 1997 *Astron. J.* **113** 1915

[10] Farago F and Laskar J 2010 *Mon. Not. Roy. Astr. Soc.* **401** 1189

[11] Doolin S and Blundell K M 2011 *Mon. Not. Roy. Astr. Soc.* **418** 2056

[12] Oks E 2015 *Astrophys. J.* **804** 106
Oks E 2016 *Astrophys. J.* **823** 69 Erratum

[13] Kryukov N and Oks E 2017 *J. Astrophys. & Aerosp. Technol.* **5** 144

[14] Oks E 2019 *New Astron.* https://doi.org/10.1016/j.newast.2019.101301

[15] Beletsky V V 2001 *Essays on the Motion of Celestial Bodies* (Basel: Birkhäuser/Springer) section 1.7

[16] Nadezhdin B B and Oks E 1986 *Sov. Tech. Phys. Lett.* **12** 512

Chapter 6

Particular analytical solution for the unrestricted three-body problem of celestial mechanics: a 'corkscrew' orbits of a planet around a binary star or of a moon around a star-planet system

The subject of this chapter is the following variation of the general three-body problem in astrophysics. There are two relatively heavy bodies rotating around their barycenter, one of the bodies being significantly heavier than the other one. In addition, there is a third, much lighter body whose orbit is relatively close to the lighter of the two heavy bodies, the orbit being *not* in the plane of the rotation of the two heavy bodies, so that the motion of the entire three-body system is really three-dimensional. One particular kind of such a system is the star–planet–moon system. Another particular kind is the binary-star–planet system (i.e., a planet around a binary star). For brevity we call such three-dimensional three-body systems '3D3BS'.

In papers [1–3] it was discovered that the third, much lighter body can have stable (rigorously speaking, metastable) orbits of a conic-helical shape. For example, in a binary-star-planet system, the planet orbit is a helix on a conical surface, whose axis of symmetry coincides with the interstellar axis (figure 6.1).

Conic-helical orbits in 3D3BS are of interest for several reasons. First, the possibility of stable (or metastable) conic-helical orbits in 3D3BS is a fundamental problem in its own right, as a relatively new chapter of the centuries-old classical three-body problem. Second, one of its applications—namely, to binary-star–planet systems—is especially significant for the search for extraterrestrial life. Indeed, according to estimates (see, e.g., papers [4–6]), approximately 50 per cent of binary stars could support habitable terrestrial planets within stable orbital ranges.

The results in papers [1–3] were obtained by applying the general analytical method for systems that can be separated into rapid and slow subsystems. The

doi:10.1088/2053-2563/ab3db0ch6

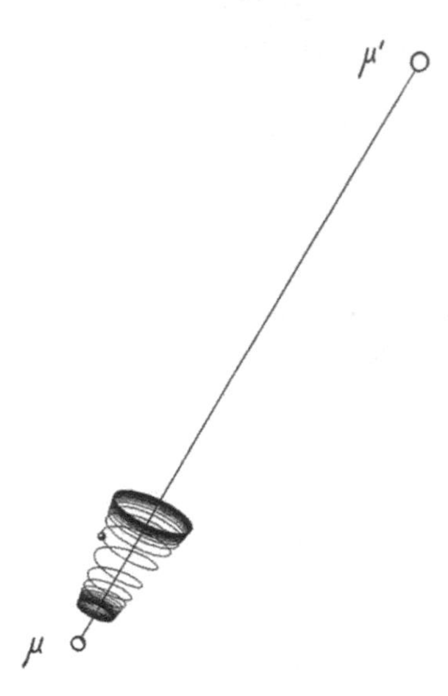

Figure 6.1. Sketch of the conic-helical motion of the planet in the model binary star system where the stars are stationary. We stretched the trajectory along the interstellar axis to make its details more visible. (Reproduced with permission from [2]. Copyright 2015 N Kryukov and E Oks.)

method was applicable because there are ranges of parameters, specified in papers [1–3], where the primary frequency Ω of the conic-helical motion of the planet in binary-star–planet systems is much greater than the Kepler frequency ω of the stars' revolution around their barycenter. Even allowing for the stars' rotation, the trajectory of the planet is still conic-helical. The plane of the quasi-circular planetary orbit undergoes relatively small oscillations along the rotating interstellar axis and the radius of the planetary orbit also undergoes relatively small oscillations. Last, but not least: in paper [1] were also obtained positive results concerning the transitability (and thus, detectability) of such planets.

The 3D3BS have a limited analogy to one-electron Rydberg quasimolecules (hereafter, ORQ) studied in atomic/molecular physics. The ORQ consist of two fully-stripped ions of the nuclear charges Z and Z' plus one highly-excited electron. They are encountered in various plasmas containing more than one kind of ions. Examples are (but are not limited to) magnetic fusion plasmas, laser-produced plasmas, plasmas used for x-ray and VUV lasers, solar plasmas, etc. In these plasmas, a fully-stripped ion of the nuclear charge Z' can come close to a hydrogen-like ion of the nuclear charge Z and form a short-lived molecule (i.e., quasimolecule). Conversely, a fully-stripped ion of the nuclear charge Z can come close to a hydrogen-like ion of the nuclear charge Z' and form a quasimolecule. Such quasimolecules are a very useful playground for theoretical and experimental studies of charge exchange, which is a physical process of primary importance for many areas of physics (e.g., for the areas listed above).

Classical analytical studies of ORQ were first presented in papers [7, 8] and later in the book [9]. The latest works were presented in the review [10]. The primary result was the discovery of classical stable electronic orbits of the shape of a helix on the surface of a cone.

It should be emphasized that there is only a limited similarity between ORQ and 3D3BS because of the following distinctions between these two physical systems. First, in ORQ the attractive centers (nuclei Z and Z') can be stationary and still be stable, while in binary star systems the rotation of the stars is necessary for the stability.

Second, in ORQ the attractive centers can engage in oscillations (called vibrations) and be stable without any rotation. This is not the case for 3D3BS.

Third, in ORQ the electronic degree of freedom has a much larger characteristic frequency and energy than the nuclear degrees of freedom. This is the basis for the standard Born–Oppenheimer approximation, where the primary contribution to the energy of the system can be obtained by freezing the nuclear motion. If necessary, the nuclear motion can be later taken into account by perturbation theory.

The Born–Oppenheimer approximation is a particular case of the general analytical method for a system that can be separated into rapid and slow subsystems. For ORQ, this method is applicable because the ratio of frequencies (and energies) of the electronic motion, the vibrational nuclear motion, and the rotational nuclear motion is 1: $(m_e/M_n)^{1/2}$: (m_e/M_n). Here m_e is the electron mass and M_n is the total mass of the two nuclei, so that $m_e/M_n < 1/3600$ and the separation of the slow and rapid subsystems is justified 'automatically'. This is, generally speaking, not the case for 3D3BS.

Indeed, for 3D3BS the analogue of molecular oscillations (vibrations) is a periodic change of the separation between the two stars, which occurs for eccentric stellar orbits. One distinction from ORQ is that both oscillations and rotations have the same frequency: the Kepler frequency ω of the two stars orbiting their barycenter. Another distinction from ORQ is that the primary frequency Ω of the helical motion of the planet is not 'automatically' much greater than the star's Kepler frequency ω.

In papers [1–3] it was shown that there are ranges of parameters where actually $\Omega \gg \omega$. Further, it was demonstrated that there are ranges of parameters where the planetary motion is stable for a model case of stationary stars and that there are ranges of parameters where the planetary motion is stable (or metastable) for the real case of stars rotating in circular or elliptical orbits. Below we reiterate the main results from papers [1–3].

First, let us consider a model system consisting of two immobile stars of masses μ and μ', and a planet of a unit mass moving around a circle in the plane perpendicular to the interstellar axis, on which the circle is centered. The mass μ is at the origin and the Oz axis is directed to the mass μ' located at $z = R$. After introducing notations

$$Z = G\mu,\ Z' = G\mu' \tag{6.1}$$

where G is the gravitational constant, the Hamilton function of the system is written in the cylindrical coordinates (z, ρ, φ) as

$$H = \frac{1}{2}\left(p_z^2 + p_\rho^2 + \frac{p_\phi^2}{\rho^2}\right) + U(z, \rho) \tag{6.2}$$

where the potential energy is

$$U(z, \rho) = -\frac{Z}{\sqrt{z^2 + \rho^2}} - \frac{Z'}{\sqrt{(R - z)^2 + \rho^2}} \tag{6.3}$$

The relation between the momenta and the corresponding velocities follows from the Hamiltonian equations of the motion:

$$\frac{dz}{dt} = \frac{\partial H}{\partial p_z} = p_z, \ \frac{d\rho}{dt} = \frac{\partial H}{\partial p_\rho} = p_\rho, \ \frac{d\phi}{dt} = \frac{\partial H}{\partial p_\phi} = \frac{p_\phi}{\rho^2} \tag{6.4}$$

Since H does not depend on φ, the corresponding momentum is conserved:

$$p_\phi = \rho^2\frac{d\phi}{dt} = M = \text{const.} \tag{6.5}$$

Physically, the separation constant M is a projection of the planet angular momentum on the interstellar axis. Thus, the z- and ρ-motions can be determined separately from the φ-motion. Then the φ-motion can be found from the ρ-motion via equation (6.5).

The Hamilton function for the z- and ρ-motions can be represented in the form

$$H = \frac{p_z^2 + p_\rho^2}{2} + U_{\text{eff}}(z, \rho) \tag{6.6}$$

where an effective potential energy (EPE) is:

$$U_{\text{eff}}(z, \rho) = \frac{M^2}{2\rho^2} + U(z, \rho) \tag{6.7}$$

After introducing scaled (dimensionless) variables w and v, a scaled projection of the angular momentum m, as well as a ratio of the star masses b,

$$w \equiv \frac{z}{R}, \ v \equiv \frac{\rho}{R}, \ m \equiv \frac{M}{\sqrt{ZR}}, \ b \equiv \frac{\mu'}{\mu} \tag{6.8}$$

the EPE can be re-written as

$$\begin{aligned} U_{\text{eff}} &= \frac{Z}{R}u_{\text{eff}}(w, v, m, b), \ u_{\text{eff}}(w, v, m, b) \\ &= \frac{m^2}{2v^2} - \frac{1}{\sqrt{w^2 + v^2}} - \frac{b}{\sqrt{(1 - w)^2 + v^2}} \end{aligned} \tag{6.9}$$

By equating to zero the derivative of the EPE with respect to w

$$\frac{\partial U_{\text{eff}}}{\partial w} = \frac{Z}{R}\frac{\partial u_{\text{eff}}}{\partial w} = 0 \tag{6.10}$$

a relation was found

$$\frac{b}{((1-w)^2 + v^2)^{3/2}} = \frac{w}{(1-w)(w^2 + v^2)^{3/2}} \tag{6.11}$$

which determines a line $v_0(w)$ in the plane (w, v), where the equilibrium points of the EPE are located:

$$v_0(w, b) = \sqrt{\frac{w^{2/3}(1-w)^{4/3} - b^{2/3}w^2}{b^{2/3} - w^{2/3}(1-w)^{-2/3}}} \tag{6.12}$$

For $b < 1$, the equilibrium value of v exists for $0 \leqslant w < b/(1+b)$ and for $1/(1+b^{1/2}) \leqslant w \leqslant 1$. For $b > 1$, the equilibrium value of v exists for $0 \leqslant w \leqslant 1/(1+b^{1/2})$ and for $b/(1+b) < w \leqslant 1$. For $b = 1$, the equilibrium value of v exists for the entire range of $0 \leqslant w \leqslant 1$. Below we refer to these intervals as the 'allowed ranges' of w.

By equating to zero the derivative of the EPE with respect to v

$$\frac{\partial U_{\text{eff}}}{\partial v} = \frac{Z}{R}\frac{\partial u_{\text{eff}}}{\partial v} = 0 \tag{6.13}$$

and then substituting $v_0(w, b)$ from equation (6.12) instead of v, another relation was found:

$$m = \frac{v_0^2(w, b)}{\sqrt{1-w}\left(w^2 + v_0^2(w, b)\right)^{3/4}} \equiv m_0(w, b) \tag{6.14}$$

While deriving equation (6.14), we used equation (6.11) to eliminate an explicit dependence on b, so that b enters equation (6.14) only implicitly—as an argument of the function $v_0(w, b)$. In a number of subsequent derivations, we will also use equation (6.11) for the same purpose without further notice.

For each set of (w, b), where w belongs to the allowed ranges, equation (6.14) determines an equilibrium value of $m_0(w, b)$—in addition to the equilibrium value of $v_0(w, b)$ determined by equation (6.12). Then for some value of b, there was considered a set of equilibrium values (w_i, v_{0i}, m_{0i}) and the EPE u_{eff} was expanded in terms of δw and δv, where

$$\delta w \equiv w - w_i,\ \delta v \equiv v - v_{0i} \tag{6.15}$$

The expansion has the form

$$u_{\text{eff}} \approx u_0 + u_{ww}\frac{\delta w^2}{2} + u_{vv}\frac{\delta v^2}{2} + u_{wv}\,\delta w\,\delta v,\ u_0 \equiv u_{\text{eff}}(w_i, v_{0i}, m_{0i}) \tag{6.16}$$

In the subsequent formulas the suffix i is dropped for brevity.

Since generally $u_{wv} \neq 0$, a rotation of the reference frame is required in order to transform the EPE to so-called 'normal' coordinates, diagonalizing the matrix of the second derivatives of the EPE [11, 12]:

$$\delta w' = \delta w \cos \alpha + \delta v \sin \alpha,\ \delta v' = -\delta w \sin \alpha + \delta v \cos \alpha \tag{6.17}$$

where

$$\mathrm{tg}2\alpha = \frac{2u_{wv}}{u_{ww} - u_{vv}} = \frac{(1 - 2w)v_0}{P} \tag{6.18}$$

so that

$$\cos \alpha = \sqrt{\frac{1 + P/Q}{2}},\ \sin \alpha = \sqrt{\frac{1 - P/Q}{2}}\,\mathrm{sign}(1 - 2w) \tag{6.19}$$

Here,

$$P \equiv w(1 - w) + v_0^2,\ Q \equiv \sqrt{\left(w^2 + v_0^2\right)\left((1 - w)^2 + v_0^2\right)} \tag{6.20}$$

In the normal coordinates, the EPE takes the form

$$u_{\mathrm{eff}} \approx u_0 + \delta w'^2 \frac{\omega_-^2}{2} + \delta v'^2 \frac{\omega_+^2}{2} \tag{6.21}$$

where

$$\omega_\pm \equiv \frac{1}{\left(w^2 + v_0^2\right)^{3/4}} \sqrt{\frac{1}{1 - w} \pm \frac{3w}{Q}} \tag{6.22}$$

The scaled (dimensionless) frequency ω_+ of small oscillations around the equilibrium in the direction of the normal coordinate $\delta v'$ is always real. According to the notations from paper [1], any frequency F and its scaled (dimensionless) counterpart f are related as follows:

$$f \equiv \sqrt{\frac{R^3}{Z}} F \tag{6.23}$$

As for the quantity ω_-, it is real if

$$v_0(w, b) \geqslant \sqrt{w(1 - w) - \frac{1}{2} + \sqrt{9w^2(1 - w)^2 - w(1 - w) + \frac{1}{4}}} \equiv v_{\mathrm{crit}}(w) \tag{6.24}$$

Physically, under the condition (6.24), the quantity ω_- is the frequency of small oscillations around the equilibrium in the direction of the normal coordinate $\delta w'$.

Thus, if $v_0(w, b) > v_{\mathrm{crit}}(w)$, the EPE has a two-dimensional minimum at the equilibrium values of w and $v = v_0(w, b)$, so that the equilibrium is stable. After introducing a scaled (dimensionless) time

$$\tau \equiv \sqrt{\frac{Z}{R^3}}\, t \tag{6.25}$$

the following final expression for the small oscillations around the stable equilibrium was obtained:

$$\begin{aligned} \delta w(\tau) &= a_w \cos(\omega_- \tau + \psi_w)\cos\alpha + a_v \cos(\omega_+ \tau + \psi_v)\sin\alpha \\ \delta v(\tau) &= a_w \cos(\omega_- \tau + \psi_w)\sin\alpha - a_v \cos(\omega_+ \tau + \psi_v)\cos\alpha \end{aligned} \tag{6.26}$$

Here amplitudes a_w, a_v and phases ψ_w, ψ_v are determined by initial conditions; $\sin\alpha$ and $\cos\alpha$ are given by equation (6.19). Compared to the corresponding equation (6.28) from [1], here in equation (6.26) we corrected a typographic error in signs.

The solution for the φ-motion turned out to be

$$\varphi(\tau) \approx f_p \tau - 2\frac{\frac{1}{\omega_-} a_w(\sin(\omega_- \tau + \psi_w) - \sin\psi_w)\sin\alpha + \frac{1}{\omega_+} a_v(\sin(\omega_+ \tau + \psi_v) - \sin\psi_v)\cos\alpha}{\sqrt{1-w}\left(w^2 + v_0^2\right)^{3/4} v_0} \tag{6.27}$$

where

$$f_p \equiv \frac{1}{\sqrt{1-w}\left(w^2 + v_0^2\right)^{3/4}} \tag{6.28}$$

is a scaled (dimensionless) primary frequency of the φ-motion. Equations (6.27) and (6.28) show that the φ-motion is a rotation about the internuclear axis with the frequency f, slightly modulated by oscillations of the scaled radius of the orbit v at the frequencies ω_+ and ω_-. In other words, for the stable motion, the planetary trajectory is a helix on the surface of a cone, with the axis coinciding with the interstellar axis. In this *conic-helical* state, the planet, while spiraling on the surface of the cone, oscillates between two end-circles which result from cutting the cone by two parallel planes perpendicular to its axis (figure 6.1).

Further, in papers [1–3], there was an analytical study of the effects of the stars' rotation and the eccentricity of their orbits on the conic-helical orbit of the planet for the situations where the Kepler frequency

$$\omega = \sqrt{\frac{G(\mu + \mu')}{R^3}} \tag{6.29}$$

of the two stars orbiting their barycenter (which is also the frequency of oscillations of the interstellar distance in the case of eccentric stellar orbits) is much smaller than the primary frequency of the revolution of the planet around the interstellar axis. This situation allows applying the standard method of the separation of rapid and slow subsystems.

According to equation (6.23), the scaled, dimensionless counterpart of the Kepler frequency is

$$f_s = \sqrt{\frac{R^3}{Z}}\omega = \sqrt{1+b} \tag{6.30}$$

The ratio of the scaled primary frequency f_p of the planetary motion (given by equation (6.28)) to the scaled Kepler frequency f_s of the stars is

$$k(w, b) = \frac{1}{\sqrt{(1+b)(1-w)}\left(w^2 + v_0^2(w, b)\right)^{3/4}} \tag{6.31}$$

This ratio becomes sufficiently large if the projection of the planetary orbit on the interstellar axis is either close to the star of the smaller mass ($w \ll 1$) or close to the star of the larger mass ($(1 - w) \ll 1$). Those are the ranges of parameters where the separation into the rapid and slow subsystems is justified.

In the reference frame rotating together with the stars with the Kepler frequency ω, there is an additional force (see, e.g., [11–13]):

$$\mathbf{F}_1 = 2\mathbf{v} \times \boldsymbol{\omega} - \boldsymbol{\omega} \times (\boldsymbol{\omega} \times \mathbf{r}) \tag{6.32}$$

Since $v \sim \Omega\rho$, where Ω is the primary frequency of the planetary motion and ρ is the average radius of the planetary orbit, and because $\Omega \gg \omega$, then the additional force is approximately

$$\mathbf{F}_1 \approx 2\mathbf{v} \times \boldsymbol{\omega} \tag{6.33}$$

Expression (6.33) has a clear physical meaning for the quantal counterpart-problem of ORQ. Namely, it is a 'Lorentz electric field' $\mathbf{v} \times \mathbf{B}_{\text{eff}}/c$ in the effective magnetic field $\mathbf{B}_{\text{eff}} = c\boldsymbol{\omega}$ in atomic units (or $\mathbf{B}_{\text{eff}} = m_e c\boldsymbol{\omega}$ in the CGS units, m_e being the electron mass).

In papers [1–3] the Ox axis was chosen along vector $\boldsymbol{\omega}$, which is obviously perpendicular to the interstellar axis chosen as the Oz axis. Since the planetary velocity $\mathbf{v}$ is primarily in the xy-plane perpendicular to the interstellar axis, then the additional force $\mathbf{F}_1$ is primarily along the interstellar axis.

Representing $\mathbf{r}/\rho = \mathbf{e}_x \cos\Omega t + \mathbf{e}_y \sin\Omega t$, so that $\mathbf{v}/\rho = (-\Omega \sin\Omega t)\mathbf{e}_x + (\Omega\cos\Omega t)\mathbf{e}_y$, and using $\boldsymbol{\omega} = \omega\mathbf{e}_x$, equation (6.32) yields:

$$\mathbf{F}_1 = \rho(-2\omega\Omega\cos\Omega t\mathbf{e_z} + \omega^2 \sin\Omega t\mathbf{e_y}) \tag{6.34}$$

In the ranges of parameters where $\Omega \gg \omega$, equation (6.34) becomes

$$\mathbf{F}_1 \approx -2\rho\omega\Omega\cos\Omega t\mathbf{e_z} \tag{6.35}$$

Thus, for the $z\rho$-motion (which in the scaled coordinates is wv-motion), the situation represents a two-dimensional oscillator, having the eigenfrequencies ω_+ and ω_- defined by equation (6.22), that is driven by the force $\mathbf{F}_1$ oscillating at the frequency Ω. Using the well-known formulas for driven oscillators (see, e.g. [13]), the solution in the coordinates w', v' rotated by the angle α (defined by equation (6.18)) compared to the coordinates w, v (i.e., in the coordinates, where the two oscillators are decoupled) can be immediately written. Then coming back to the original scaled coordinates w, v, one obtains:

$$\delta w(\tau) = 2v_0(w, b)\omega_s f_p \left(\frac{\cos^2 \alpha}{f_p^2 - \omega_-^2} - \frac{\sin^2 \alpha}{f_p^2 - \omega_+^2} \right) \cos f_p \tau$$
$$\delta v(\tau) = 2v_0(w, b)\omega_s f_p \sin \alpha \cos \alpha \left(\frac{1}{f_p^2 - \omega_-^2} + \frac{1}{f_p^2 - \omega_+^2} \right) \cos f_p \tau \tag{6.36}$$

Using the relation $(\omega_+^2 + \omega_-^2)/2 = f_p^2$ and equation (6.19), the latter formulas can be finally re-written in the following simple form:

$$\delta w(\tau) = \frac{4v_0(w, b)\omega_s f_p}{\omega_+^2 - \omega_-^2} \cos f_p \tau$$
$$\delta v(\tau) = 0 \tag{6.37}$$

We note that in equations (6.32)–(6.37), we corrected some typographic errors compared to the corresponding equations from paper [1].

Equation (6.37) shows, in particular, that the major term in the amplitude $\delta v(\tau)$ of the oscillations of the scaled radius of the planetary orbit—the term that could have been of the same order as the right side of $\delta w(\tau)$—vanished. Physically, this means that $\delta v(\tau) \ll \delta w(\tau)$, where a small non-zero contribution to $\delta v(\tau)$ could be obtained by taking into account the second, small term in equation (6.34). Thus, while the trajectory of the planet is conic-helical, the cone is very close to a cylinder. In other words, the forced oscillations of the 'circular' planetary orbit are primarily along the interstellar axis. For the quantal counterpart-problem of ORQ this should have been expected. Indeed, the electron in a circular orbit is like a charged 'ring': so, in the monochromatic electric (Lorentz) field perpendicular to the axis of the charged ring, the ring should oscillate along its axis.

Figure 6.2 shows the scaled amplitude $\delta w_0 = 4v_0(w, b)\omega_s f_p/(\omega_+^2 - \omega_-^2)$ of the oscillations of the planetary orbit along the interstellar axis versus the scaled

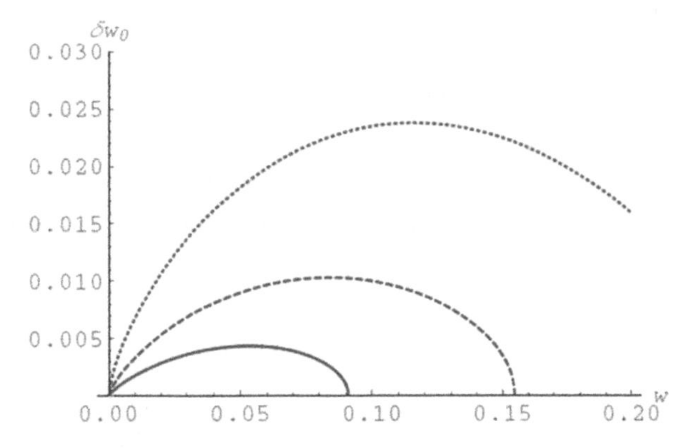

Figure 6.2. The scaled amplitude δw_0 of the oscillations of the planetary orbit along the interstellar axis versus the scaled projection $w = z/R$ of the average plane of the planetary orbit on the interstellar axis, for three values of the ratio b of the stellar masses: b = 100 (solid line), b = 30 (dashed line), and b = 10 (dotted line). (Reproduced with permission from [2]. Copyright 2015 N Kryukov and E Oks.)

projection $w = z/R$ of the average plane of the planetary orbit on the interstellar axis, for three values of the ratio b of the stellar masses. It is seen that $\delta w_0 \ll 1$ (i.e., the amplitude of the oscillations is much smaller than the interstellar distance) for b greater or of the order of 10. For the validity condition $\Omega \gg \omega$ of this result to be satisfied with a large margin of 'safety', the average plane of the planetary orbit should be very close to the star of the smaller mass (and the closer it is in the range of $w < 0.01$, the smaller is the amplitude δw_0, as seen from figure 6.2). As long as the primary frequency Ω of the planet revolution about the interstellar axis exceeds by many orders of magnitude the Kepler frequency ω of the stars' rotation, the conic-helical planetary orbit would remain stable for a very long time. (Rigorously speaking, the planet is in a metastable state, which is yet another analogy with atomic/molecular systems: they have metastable states living by many orders of magnitude longer than other states of the system.)

Finally, the effect of the eccentric orbits of the stars was studied in papers [1–3]. In the reference frame rotating together with the stars with the Kepler frequency ω, a non-zero eccentricity ε of the stars' orbits results in the oscillation of the interstellar distance R with the frequency ω. In the ranges of parameters where $\Omega \gg \omega$, the oscillation of the interstellar distance is an adiabatic 'perturbation' of the planetary motion. According to the principle of adiabatic invariance, the planetary motion will adjust to the slowly varying R while keeping as the constant the average nonzero projection M of the planetary angular momentum on the interstellar axis. The projection of the planetary angular momentum on the interstellar axis undergoes small oscillations (caused by stars' rotation) around the nonzero average M, the latter being the adiabatic invariant. Particularly, for the average plane of the planetary orbit close to the star of the smaller mass, it was shown in papers [1–3] that the eccentricity ε of the stars' orbit would not affect the stability of the planetary motion if ε does not exceed some critical value $\varepsilon_c(b)$.

Now we discuss the same problem in the relativistic framework following papers [2, 3]. The relativistic force acting on the planet (scaled by its mass, i.e., the force divided by the mass of the planet) is given by the formula

$$\mathbf{F} = \gamma_0\left(\mathbf{a} + \mathbf{V}\frac{\gamma_0^2}{c^2}\mathbf{V}\cdot\dot{\mathbf{V}}\right) \tag{6.38}$$

where $\mathbf{V}$ is the velocity of the planet and $\gamma_0 = (1 - V^2/c^2)^{-1/2}$. The additional acceleration given by equation (6.32) can be substituted into equation (6.38) as the last factor, with $\mathbf{V}$ as the time derivative of the position in the rotating reference frame

$$\mathbf{r} = \rho\cos\Omega t\mathbf{e}_x + \rho\sin\Omega t\mathbf{e}_y \tag{6.39}$$

After the calculation, we obtain the following values for the components of the force:

$$F_x = -\gamma_0 \omega V \frac{\gamma_0^2 \beta^2}{2} \frac{\omega}{\Omega} \sin 2\Omega t \sin \Omega t$$
$$F_y = \gamma_0 \omega V \frac{\omega}{\Omega} \left(\sin \Omega t + \frac{\gamma_0^2 \beta^2}{2} \sin 2\Omega t \cos \Omega t \right) \tag{6.40}$$
$$F_z = -2\gamma_0 \omega V \cos \Omega t$$

(here $\beta = V/c$). As $\omega \ll \Omega$, the dominant term is the z-projection of the force (given by the third equation in (6.40)). By analogy with the non-relativistic case, we derive the small oscillations about the equilibrium due to the dominant force term:

$$\delta w(\tau) = \frac{4v\tilde{\omega}\tilde{\Omega}}{({\omega_+}^2 - {\omega_-}^2)\sqrt{1 - \beta^2}} \cos \tilde{\Omega}\tau,\ \delta v(\tau) = 0 \tag{6.41}$$

where $v = \rho/R$ (its equilibrium value given by equation (6.12)) and the tilde above means scaling by multiplying by $(R^3/Z)^{1/2}$. We can now find the conditions, under which the amplitude of the oscillations is small while the condition $\omega \ll \Omega$ stays valid. In the relativistic case, the Hamiltonian, divided by the mass m of the planet $h = H/m$, for the 3D3BS case is given by

$$h = c\sqrt{m^2c^2 + p_z^2 + p_\rho^2 + \frac{p_\phi^2}{\rho^2}} - \frac{Z}{\sqrt{z^2 + \rho^2}} - \frac{Z'}{\sqrt{(R - z)^2 + \rho^2}} - mc^2 \tag{6.42}$$

In a circular state, $p_z = p_\rho = 0$ and $|p_\varphi|/m = \text{const.} = L$. Using the scaling

$$\ell = \frac{L}{cR},\ \varepsilon = -\frac{R}{Z}E,\ r = \frac{Z}{L^2}R \tag{6.43}$$

we write the scaled energy for the circular state:

$$\varepsilon = \frac{1}{\sqrt{w^2 + v^2}} + \frac{b}{\sqrt{(1 - w)^2 + v^2}} + \frac{c^2R}{Z}\left(1 - \sqrt{1 + \frac{\ell^2}{v^2}}\right) \tag{6.44}$$

For the relativistic motion,

$$L = \frac{mV\rho}{\sqrt{1 - \frac{V^2}{c^2}}} \tag{6.45}$$

and, using the scaling $\rho = vR$ from equation (6.8) and the first formula in equation (6.43), we can find the speed of the planet in the relativistic OBSS case in the units of the speed of light:

$$\beta = \frac{1}{\sqrt{1 + \frac{p}{\ell^2}}} \tag{6.46}$$

where $p = v^2$. From the first and the third formulas in equation (6.43), $\ell^2 = \alpha/r^2$, where $\alpha = (Z/(cL))^2$. Thus,

$$\beta = \frac{1}{\sqrt{1 + \frac{pr^2}{\alpha}}} \tag{6.47}$$

The equilibrium values for p and r can be obtained from differentiating (6.44) with respect to w and v. The first differentiation gives the same relation as in the non-relativistic case, so the equilibrium value of p is the squared right-hand side of equation (6.12). The second differentiation (with respect to v), with the later substitution of the equilibrium value of v (or p), yields the equilibrium value of ℓ, which is related to r by $\ell^2 = \alpha/r^2$ mentioned above. Using the substitution

$$\gamma = \left(\frac{1}{w} - 1\right)^{1/3} \tag{6.48}$$

which significantly simplifies the formulas in the two-Coulomb-center problem, we find the speed of the planet in the circular state (in the units of the speed of light):

$$\beta = \sqrt{\frac{\alpha(\gamma^4 - b^{2/3})^3}{(\gamma^3 - 1)^3(\gamma^3 + 1)}} \tag{6.49}$$

From equation (6.45) with the substituted value of $V = \Omega\rho$, the first relation in equation (6.43), the second relation in equation (6.8) and using $\ell^2 = \alpha/r^2$, we find

$$\Omega = \frac{c}{R}\frac{\sqrt{\alpha}}{pr}\sqrt{1 - \beta^2} \tag{6.50}$$

and

$$\tilde{\Omega} = c\sqrt{\frac{R}{Z}}\frac{\sqrt{\alpha}}{pr}\sqrt{1 - \beta^2} \tag{6.51}$$

As given in equation (6.1), and measuring now the mass of the star μ in the units of the mass of the Sun (in distinction to the nonrelativistic case, where the star masses were measured in units of the mass of the planet), we get

$$Z = GM_{\odot}\mu \tag{6.52}$$

where $M_{\odot} = 1.989 \times 10^{33}$ g. Substituting equation (6.52) into equation (6.51), we obtain

$$\tilde{\Omega} = \sqrt{\frac{R}{\mu s}}\frac{\sqrt{\alpha}}{pr}\sqrt{1 - \beta^2} \tag{6.53}$$

where the quantity $s = G\,M_{\odot}/c^2$ (which is one half of the Schwarzschild radius of the Sun and is approximately equal to 147 700 cm).

As defined above, $\alpha^{1/2} = Z/(cL)$, and substituting the scaling relation for L from equation (6.43), $\ell^2 = \alpha/r^2$ and equation (6.52), we have

$$\alpha = \frac{\mu s}{R} r \tag{6.54}$$

We substitute equation (6.54) into equation (6.49), and then substitute the resulting equation for β and the solution for α into equation (6.53), obtaining the equation for the scaled frequency of the revolution of the planet that only depends on the ratio $R/(\mu s)$ and γ (or w—see equation (6.48)). Finally, from equation (6.41), and because $\tilde{\omega} = (1 + b)^{1/2}$ from equations. (6.29) and (6.30), the amplitude of the small oscillations about the equilibrium on the w-axis is

$$\delta w_0 = \frac{4\sqrt{p}\sqrt{1+b}}{\left|\omega_+{}^2 - \omega_-{}^2\right|\sqrt{1-\beta^2}}\tilde{\Omega} \tag{6.55}$$

and substituting the equation for the scaled frequency of the revolution of the planet obtained previously, the equilibrium values for p and r, the equation for β and the frequencies from equation (6.22), we derive the amplitude of the small oscillations of the planet on the w-axis in the relativistic case:

$$\delta w_0 = \frac{2b^{1/3}\sqrt{1+b}\sqrt{\gamma}(\gamma^3 - 1)^{7/4}\sqrt{\gamma^4 - b^{2/3}}}{3(b^{2/3}\gamma^2 - 1)^{9/4}(\gamma^3 + 1)^{5/4}} \tag{6.56}$$

The amplitude in the relativistic case is the same as in the non-relativistic case.

We checked that the ratio k of the frequencies of the revolution of the planet Ω and the Kepler frequency ω of the revolution of the stars about their barycenter is much greater than 1. By using the previously found values of Ω and ω and substituting the equilibrium values of p, r, α and β, we derive the formula for the ratio k of the frequencies depending on the axial coordinate for the given interstellar distance R and the star masses μ and μ':

$$k = \frac{(b^{2/3}\gamma^2 - 1)^{3/4}(\gamma^3 + 1)^{5/4}}{\sqrt{1+b}\,\gamma^{3/2}(\gamma^3 - 1)^{3/4}}\sqrt{1 - \frac{\mu s}{R}\frac{(\gamma^4 - b^{2/3})\sqrt{(b^{2/3}\gamma^2 - 1)(\gamma^3 + 1)}}{\gamma(\gamma^3 - 1)^{3/2}}} \tag{6.57}$$

From equation (6.57) it can be found out that k in the relativistic case is equal to the non-relativistic k_{NR} multiplied by $(1 - \beta^2)^{1/2}$. Figure 6.3 shows the ratio k/k_{NR} versus the scaled radius of the orbit v, for the masses of the stars $\mu = 1$ and $\mu' = 100$ (in the units of the mass of the Sun) and the interstellar distance $R = 100$ a.u. It is seen that the relativistic effects become significant when $v \sim 10^{-9}$ or smaller.

The range of $v = (1 \div 2) \times 10^{-9}$ for $R = 100$ a.u. corresponds to the range of the radius of the planetary orbit $\rho = (15 \div 30)$ km. Since the radius of the planet should be smaller than ρ, in this case it should be a planetoid.

Figure 6.4 shows the dependence $k(v)$ for the example where the mass of the lighter star $\mu = 1$ and the heavier star $\mu' = 100$ (in the units of the mass of the Sun) and for the interstellar distance $R = 100$ a.u. (wide binary system). It is seen that the

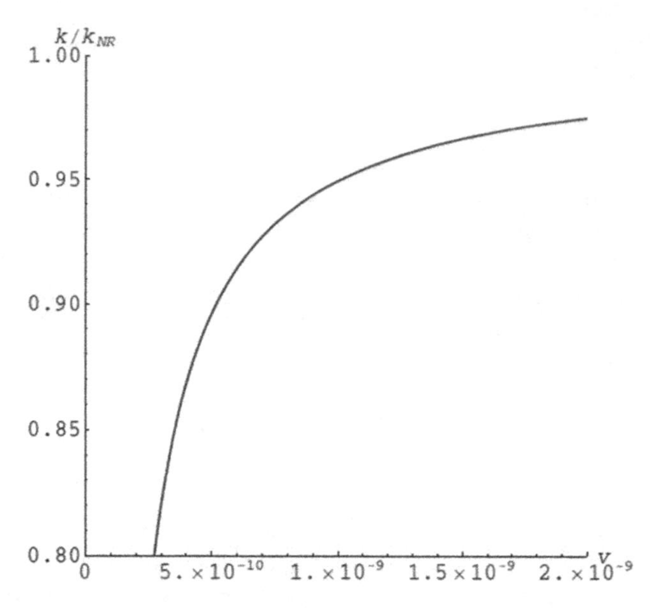

Figure 6.3. The ratio of the relativistic and non-relativistic k (which is the ratio of the frequencies of the revolution of the planet Ω and the Kepler frequency ω of the revolution of the stars about their barycenter) versus the scaled radius of the planetary orbit $v = \rho/R$, for the masses of the stars $\mu = 1$ and $\mu' = 100$ (in the units of the mass of the Sun) and the interstellar distance $R = 100$ a.u. (Reproduced with permission from [2]. Copyright 2015 N Kryukov and E Oks.)

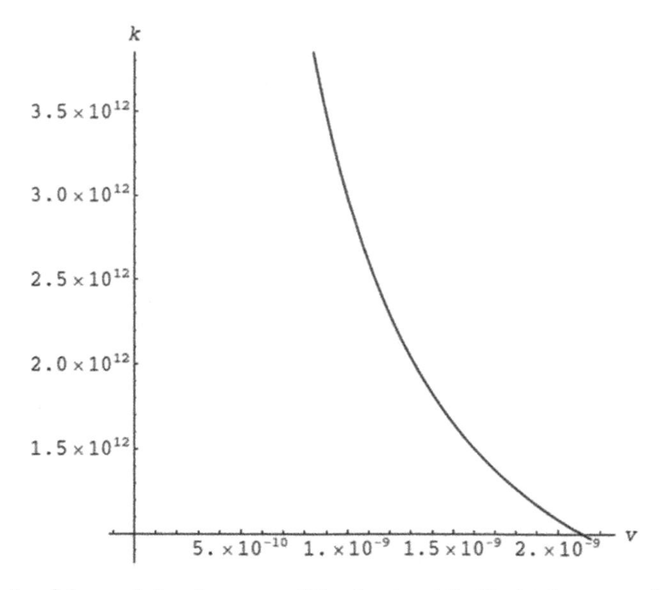

Figure 6.4. The ratio of the revolution frequency of the planet and the Kepler frequency of the rotation of the stars versus the scaled orbit radius $v = \rho/R$, for the example where the mass of the lighter star $\mu = 1$ and the heavier star $\mu' = 100$ (in the units of the mass of the Sun) and the interstellar distance $R = 100$ a.u. in the relativistic case. (Reproduced with permission from [2]. Copyright 2015 N Kryukov and E Oks.)

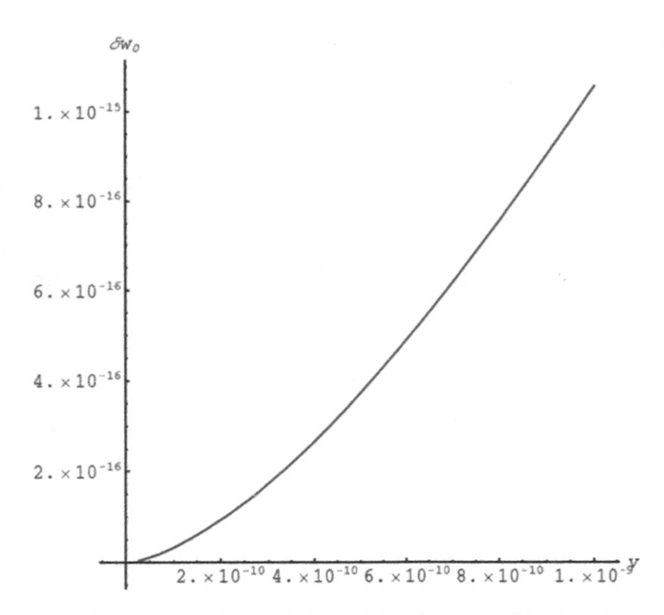

Figure 6.5. The scaled amplitude δw_0 of the oscillations of the planetary orbit along the interstellar axis versus the scaled orbit radius $v = \rho/R$ for the example where the mass of the lighter star $\mu = 1$ and the heavier star $\mu' = 100$ (in the units of the mass of the Sun). (Reproduced with permission from [2]. Copyright 2015 N Kryukov and E Oks.)

ratio of the frequencies is greater than 10^{12} for $v < 2 \times 10^{-9}$, which means that such a system would be stable for a very long time.

In figure 6.5, we present the plot of the scaled amplitude δw_0 of the axial oscillations of the planetary orbit from equation (6.56) versus the scaled radial coordinate v for the mass of the lighter star $\mu = 1$ and the heavier star $\mu' = 100$, in the units of the mass of the Sun. It is seen that the amplitude of the oscillations is much (about one million times) smaller than the interstellar distance.

At the end of this chapter, we briefly discuss another example of the 3D3BS: a star–planet–moon system [3]. We consider, as an example, an Earth-like planet around a Sun-like star, so that the ratio of the masses μ/μ' of the planet and the star is 3×10^{-6}. The separation between the planet and the star is $R = 1$ AU. The planet has a moon.

In paper [3] it was shown analytically that the ratio Ω/ω of the frequency Ω of the moon revolution around the planet–star axis to the frequency ω of the planet rotation around the star is much greater than unity for the scaled projection $w = z/R$ of the average plane of the moon's orbit on the planet-star axis in the range of $0 < w < 10^{-4}$. So, the separation of rapid and slow subsystems was again justified.

It was also demonstrated in paper [3] that the scaled amplitude δw_0 of the oscillations of the moon's orbit along the planet–star axis versus the scaled coordinate $w = z/R$ can be very small: $\delta w_0 \sim 10^{-6}$ at $w < 10^{-4}$. Further, in paper [3] it was demonstrated that the scaled amplitude δv of the oscillations of the scaled radius $v = \rho/R$ of the moon's orbit in the plane perpendicular to the planet–star axis versus the scaled coordinate w can also be very small:

$\delta v \sim 10^{-6}$ at $w < 10^{-4}$.

In addition to the above analytical results, the authors of paper [3] also performed exact simulations of the moon's motion, as the moon revolves around the planet–star axis while this axis rotates with the Kepler frequency corresponding to the star–planet two-body problem. The simulations were performed for the initial value of the scaled projection of the average plane of the moon's orbit on the planet–star axis $w_0 = 10^{-4}$, i.e., 15 000 km. It was found that all the three coordinates of the moon underwent only relatively small oscillations around their equilibrium values, so that the conic-helical orbit of the moon in the star–planet–moon system is stable.

Thus, under the influence of a relatively distant star, the plane of the moon's orbit could orient itself to be perpendicular to the planet–star axis. The trajectory of the moon becomes conic-helical. This also means that the standard analytical method of separating rapid and slow subsystems, used by the authors of paper [3], is well justified and that this configuration will remain stable for a very long time.

References

[1] Oks E 2015 *Astrophys. J.* **804** 106
Oks E 2016 *Astrophys. J.* **823** 69

[2] Krjukov N and Oks E 2015 *Internat. Rev. Atom. Mol. Phys.* **6** 7

[3] Kryukov N and Oks E 2017 *J. Astrophys. Aerosp. Technol.* **5** 144

[4] Quintana E V and Lissauer J J *In Planets in Binary Star Systems* ed N Naghighipour (Dordrecht: Springer), ch 10 265

[5] Fatuzzo M, Adams F C, Gauvin R and Proszkow E M 2006 *Pub. Astron. Soc. Pacific* **118** 1510

[6] David E, Quintana E V, Fatuzzo M and Adams F C 2003 *Pub. Astron. Soc. Pacific* **115** 825

[7] Oks E 2000 *Phys. Rev. Lett.* **85** 2084

[8] Oks E 2000 *J. Phys. B: Atom. Mol. Opt. Phys.* **33** 3319

[9] Oks E 2006 *Stark Broadening of Hydrogen and Hydrogenlike Spectral Lines in Plasmas: The Physical Insight* (Oxford: Alpha Science International) appendix A

[10] Kryukov N and Oks E 2013 *Inter. Rev. Atom. Mol. Phys.* **4** 121

[11] Landau L D and Lifshitz E M 1960 *Mechanics* (Oxford: Pergamon)

[12] Goldstein H 1980 *Classical Mechanics* (Reading, MA: Addison-Wesley)

[13] Jose J V and Saletan E J 1998 *Classical Dynamics: A Contemporary Approach* (Cambridge: Cambridge University Press) section 4.2.4

Chapter 7

Magnetic stabilization of one-electron Rydberg quasimolecules

In paper [1], the authors obtained an exact analytical classical solution for the electronic terms of circular Rydberg states (CRS) in the presence of a magnetic field B for two-Coulomb-center systems. The classical electronic terms were shown to be significantly affected by the magnetic field. In particular, a sufficiently strong magnetic field is shown to cause the appearance of CRS above the ionization threshold. These CRS are the classical molecular counterparts of the quantal atomic quasi-Landau levels (resonances).

In paper [1], the authors focused on the analytical classical description of CRS of two-Coulomb-center systems in a magnetic field B parallel to the internuclear axis. The system consists of two nuclei of charges Ze and $Z'e$, separated by a distance R, and one electron, and is denoted by ZeZ'. Analytical results for the electronic terms $E(R)$ of the ZeZ'-system for the field-free case were obtained [2, 3] from first principles within a purely classical approach. The classical approach reproduces [2, 3] several electronic terms and two of these terms undergo a V-shape crossing at separation R^*, so that CRS cannot exist for $R < R^*$.

In paper [1] the authors obtained an exact analytical classical solution for the electronic terms $E(R,B)$ for CRS of the ZeZ'-system in the presence of a magnetic field B. The solution is exact and is valid for any strength of the magnetic field. They also studied how the classical electronic terms are influenced by the magnetic field, including the case of a strong field. This is a fundamental problem in its own right.

Then they used the theory to explore the stability of the nuclear motion in the ZeZ'-system. According to the method of separating rapid (electronic) and slow (nuclear) subsystems, the electronic term $E(R,B)$ should be added to the potential for the nuclear motion. It was found that the term $E(R,B)$ in the effective potential $v(R,B) = ZZ'/R + E(R,B)$ for the relative motion of the nuclei plays a crucial role. They found that the CRS-system, in the absence of the magnetic field, is not a stable molecule, but is only a quasi-molecule with anti-bonding molecular orbitals.

doi:10.1088/2053-2563/ab3db0ch7

A similar classical result was obtained by Pauli [4] for the hydrogen molecular-ion H_2^+. The authors of paper [1] found that a magnetic field creates a deep minimum in one of the branches of the effective potential $V(R,B)$ for relative motion of the nuclei, so as to render stable nuclear motion. The magnetic field can therefore be used to transform a quasi-molecule into a real CRS molecule with a bonding molecular orbital. This finding initiates a new phenomenon—*the magnetically-controlled stabilization of the CRS quasi-molecules*—suitable for further studies. Let the charge Z of the two-Coulomb-center system be fixed at the origin and the charge Z' be located along the OZ axis at nuclear separation R. For simplicity, let the plane of the electron's circular orbit of radius ρ centered at z be perpendicular to the internuclear axis OZ. For $z \ll R$ or for $(R - z) \ll R$ when the electron is mainly bound to the Z or the Z' ion and is perturbed by the other fully stripped ion, these circular orbits depict Stark states which correspond classically to zero projection of the Runge–Lenz vector [5] on the axis OZ and quantally to zero electric quantum number $k = n_1 - n_2$, where n_1 and n_2 are the parabolic quantum numbers [6]. The classical Hamiltonian for fixed R of the ZeZ'-system in the presence of a uniform magnetic field B parallel to the internuclear axis is given in *atomic units* by

$$H(\rho, z) = M^2/(2\rho^2) - Z/(\rho^2 + z^2)^{1/2} - Z'/[\rho^2 + (z - R)^2]^{1/2} \\ +\Omega M + \Omega^2\rho^2/2, \;\; \Omega \equiv B/(2c). \tag{7.1}$$

Here M is the constant z-component of the angular momentum and Ω is the Larmor frequency expressed in practical units as $\Omega(s^{-1}) \approx 8.794 \times 10^6\, B(G)$.

Let us introduce the following scaled quantities:

$$b \equiv Z'/Z, \quad u \equiv \rho/R, \quad w \equiv z/R, \quad m \equiv M/(ZR)^{1/2}, \\ \omega \equiv WM^3/Z^2, \quad h \equiv HM^2/Z^2, \tag{7.2}$$

so that the scaled Hamiltonian is

$$h(u, w, \omega) = m^2\varepsilon(u, w, \omega), \\ \varepsilon(u, w, \omega) \equiv m^2/(2u^2) - 1/(u^2 + w^2)^{1/2} - b/[u^2 + (1 - w)^2]^{1/2} \\ +\omega/m^2 + \omega^2u^2/(2m^6). \tag{7.3}$$

The conditions for dynamic equilibrium are,

$$\partial h/\partial w = m^2\{w/(u^2 + w^2)^{3/2} - b(1 - w)/[u^2 + (1 - w)^2]^{3/2}\} = 0, \tag{7.4}$$

and

$$\partial h/\partial u = m^2\{-m^2/u^3 + u/(u^2 + w^2)^{3/2} + bu/[u^2 + (1 - w)^2]^{3/2} + \omega^2u/m^6\} = 0. \tag{7.5}$$

Equation (7.4) shows that equilibrium along the internuclear axis does not depend on the scaled magnetic field ω. In terms of the equilibrium value w_0 of w, the equilibrium value of u can therefore be expressed as,

$$u(w_0, b) = \left\{[w_0(1 - w_0)^2]^{2/3} - b^{2/3}w_0^2\right\}^{1/2} / \{b^{2/3} - [w_0/(1 - w_0)]^{2/3}\}^{1/2}. \tag{7.6}$$

which only exists within the following 'allowed ranges',

$$\begin{aligned}
&0 \leqslant w_0 < b/(1+b) \text{ and } 1/(1+b^{1/2}) \leqslant w_0 \leqslant 1; \quad b < 1; \\
&0 \leqslant w_0 \leqslant 1/(1+b^{1/2}) \text{ and } b/(1+b) < w_0 \leqslant 1; \quad b > 1; \\
&0 \leqslant w_0 \leqslant 1; \qquad\qquad\qquad\qquad\qquad\qquad b = 1;
\end{aligned} \tag{7.7}$$

of w_0. Equation (7.5), represents the condition for equilibrium in the orbital plane and can be re-written in the form

$$\begin{aligned}
&m(w_0, b, \omega)= \\
&\pm\{f/4 + (f^2/4 + j)^{1/2}/2 + [f^2/2 - j + (f^3/4)/(f^2/4 + j)^{1/2}]^{1/2}/2\}^{1/2},
\end{aligned} \tag{7.8}$$

where, in terms of $u(w_0, b)$, given by equation (7.6),

$$\begin{aligned}
f(w_0, b, \omega) \equiv\, & u^4(w_0, b)/[u^2(w_0, b) + w_0{}^2]^{3/2} \\
& + bu^4(w_0, b)/[u^2(w_0, b) + (1 - w_0)^2]^{3/2},
\end{aligned} \tag{7.9}$$

and

$$j(w_0, b, \omega) \equiv [u^4(w_0, b)\omega^2/18]^{1/3}g - (4/g)[2u^8(w_0, b)\omega^4/3]^{1/3}, \tag{7.10}$$

with

$$g(w_0, b, \omega) \equiv |-9f^2 + [81f^4 + 768u^4(w_0, b)\omega^2]^{1/2}|^{1/3}. \tag{7.11}$$

The plus and minus signs in equation (7.8) correspond, respectively to the positive and negative projections of the angular momentum along the magnetic field. For each set $\{b, m, \omega\}$ of parameters, equation (7.8) determines the equilibrium value w_0 of the scaled z-coordinate of the orbital plane.

The internuclear distance R as noted above was considered to be 'frozen'. In order to reproduce the electronic terms, i.e., the dependence of the electronic energy on the internuclear distance, one should now allow R to be a slowly varying adiabatic quantity (slowly varying with respect to the electronic motion, as in the Born–Oppenheimer approach [7]).

We consider energy terms of the *same symmetry* which, for the quantal ZeZ'-problem, means terms with the same magnetic quantum number M [8–12]. Therefore, in our classical ZeZ'-problem, from now on we consider fixed projection of the angular momentum M and study the behavior of the classical energy keeping M constant.

We introduce the scaled internuclear distance

$$r \equiv RZ/M^2 \tag{7.12}$$

which, under the fourth relation in equation (7.2), reduces to

$$r(w_0, b, \omega) = 1/m^2(w_0, b, \omega). \tag{7.13}$$

On substituting $w = w_0$ into equation (7.3), then,

$$h(w_0, b, \omega) = m^2(w_0, b, \omega)\varepsilon[u(w_0, b), w_0, \omega]. \tag{7.14}$$

Thus, for any positive ratio of the nuclear charges $b > 0$ and for any value of the scaled magnetic field ω, the dependence of the scaled energy h on the scaled internuclear distance r is determined by equations (7.13) and (7.14) in terms of the parameter w_0, which takes all values from the allowed ranges specified by equation (7.7) i.e., equations (7.13), (7.14) determine the *classical electronic energy terms* for any strength of the magnetic field, including the strong field case. Particular examples in paper [1] were chosen for the one electron quasimolecule based on nuclei of He and Li: $Z = 2$ and $Z' = 3$, so that $b = 3/2$. In the present chapter, for demonstrating that the qualitative results of paper [1] hold not only for $Z = 2$ and $Z' = 3$, we choose for examples the one-electron quasimolecule based on nuclei of He and Be: $Z = 2$ and $Z' = 4$, so that $b = 2$. Figure 7.1 shows the scaled electronic energy h versus the scaled internuclear distance r for $b = 2$ in the absence of magnetic field. There are three terms of the same symmetry, a totally counterintuitive result because there is more than one classical energy term. Moreover, two of these classical energy terms undergo a V-shape crossing.

We note that the upper and middle energy terms terminate at some $r = r_{min}$, so that there are no CRS at $r < r_{min}$ for these two energy terms. The classical energy of the CRS acquires an imaginary part at $r < r_{min}$, corresponding quantally to virtual states/resonances. There may well be non-circular Rydberg states at $r < r_{min}$ in the same energy range, but this was beyond the scope paper [1].

We emphasize that the above example of $Z'/Z = 3/2$ is fully representative. In fact, for any pair of Z and $Z' \neq Z$, there are three classical energy terms of the same symmetry and the upper term always crosses the middle term. (For $Z' = Z$, there is only one term in the corresponding plot and no crossing, as expected.)

The previously published analysis [2, 3] provided the following interpretation of these three energy terms. The lower term, as $R \to \infty$ corresponds to the energy $E \to -(Z_{max}/M)^2/2$ of the hydrogen-like ion with nuclear charge $Z_{max} \equiv \max(Z', Z)$ perturbed slightly by the other charge $Z_{min} \equiv \min(Z', Z)$. As $R \to 0$, the lower term

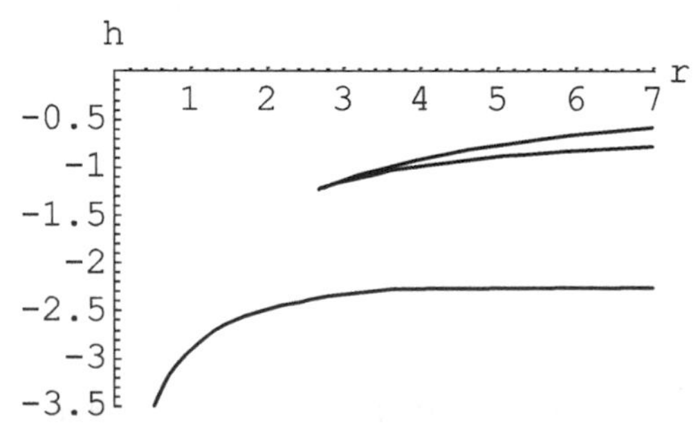

Figure 7.1. The scaled electronic energy h versus the scaled internuclear distance r for the ratio of the nuclear charges $b = 2$ at the absence of the magnetic field (h and r are defined by equations (7.2) and (7.12), respectively).

translates into the energy $E \to (-(Z + Z')^2/M)^2/2$ of the hydrogen-like ion of the nuclear charge $Z + Z'$, i.e., to the united–atom limit [8–12]. The middle term as $R \to \infty$ corresponds to the energy $E \to -(Z_{min}/M)^2/2$ of the hydrogen-like ion of nuclear charge Z_{min} slightly perturbed by the charge Z_{max}. The upper term, as $R \to \infty$, evolves into a near-zero-energy state.

The analysis presented in paper [3] was not confined to circular orbits of the electron. In order to make the present work more transparent, we briefly outline here the scheme of that analysis. In cylindrical coordinates (z, ρ, ϕ), using the axial symmetry of the problem, the z- and ρ-motions, due to axial symmetry, can be separated from the ϕ-motion. The ϕ-motion can be determined from the calculated ρ-motion. Equilibrium points of the two-dimensional motion in the $z\rho$-space were studied and a condition distinguishing between two physically different cases where the effective potential energy: (a) has a two-dimensional minimum in the $z\rho$-space, and (b) has a saddle point in the $z\rho$-space was explicitly derived. In particular, it turned out that the boundary between these two cases corresponds to the point of crossing of the upper and middle energy terms. For stable motion, the trajectory was found [3] to be a helix on the surface of a cone, with axis coinciding with the internuclear axis. In this *helical* state, the electron, while spiraling on the surface of the cone, oscillates between two end-circles which result from cutting the cone by two parallel planes perpendicular to its axis see figure 6.1.

We now 'turn on' the magnetic field—in contrast to the scope of papers [2, 3]. Figure 7.2 shows the scaled electronic energy h versus the scaled internuclear distance r for $b = 2$ at $\omega = +1.18$, i.e., at a moderate value of the magnetic field. We note that $\omega > 0$ corresponds to $BM > 0$, while $\omega < 0$ corresponds to $BM < 0$; remember B and M are the z-projections of the magnetic field and of the angular momentum, respectively, and that the Oz axis is directed from the charge Z toward the charge Z'.

Figure 7.2 shows that the magnetic field corresponding to $\omega = +1.18$ and higher values, under the condition $BM > 0$, lifts the entire upper and middle energy terms into the continuum. Figure 7.3 shows the scaled electronic energy h versus the scaled internuclear distance r for $b = 2$ at $\omega = +3.8$, i.e., at a larger value of the magnetic

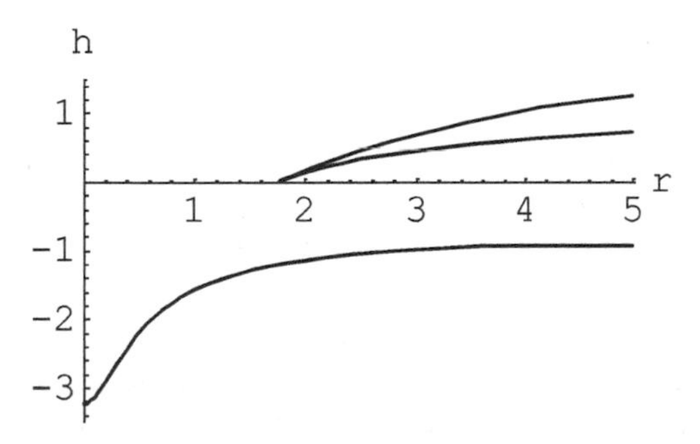

Figure 7.2. Same as in figure 7.1, but at the scaled magnetic field $\omega = +1.18$. We note that $\omega > 0$ corresponds to $BM > 0$, while $\omega < 0$ corresponds to $BM < 0$; here B and M are z-projections of the magnetic field and of the angular momentum, respectively; the Oz axis is directed from the charge Z toward the charge Z'.

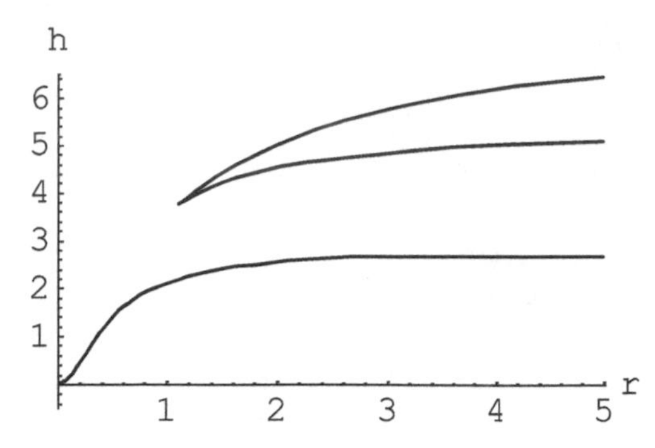

Figure 7.3. Same as in figure 7.2, but at $\omega = +3.8$.

field. It is seen that the magnetic field of this value (and of higher values), under the condition $BM > 0$, lifts all three energy terms into the continuum.

These CRS above the ionization threshold, shown in figures 7.2 and 7.3, are *classical molecular counterparts of the quantal atomic quasi-Landau levels* or resonances. The latter were discovered experimentally by Garton and Tomkins [13] (for theoretical references on atomic quasi-Landau resonances, see, e.g., the book [14]).

Now we study the stability of the nuclear motion in the ZeZ'-system. As noted above, the electronic term $E(R,B)$ should be added to the potential for the nuclear motion—according to the method of separating rapid (electronic) and slow (nuclear) subsystems—see, e.g., book [15]. The electronic energy $E(R, B)$ becomes a crucial part of the effective internuclear potential

$$V(R, B) = ZZ'/R + E(R, B). \tag{7.15}$$

for the relative motion of the nuclei. The scaled internuclear potential

$$v \equiv VM^2/Z^2, \tag{7.16}$$

then reduces (cf equation (7.14)) to

$$v(w_0, b, Z, \omega) = m^2(w_0, b, \omega)\{\varepsilon[u(w_0, b), w_0, \omega] + Z'\}. \tag{7.17}$$

For any set $\{b, Z', \omega\}$, equations (7.13) and (7.17) therefore determine the dependence of the scaled internuclear potential v on the scaled internuclear distance r in terms of the parameter w_0 which takes all values within the allowed ranges specified by equation (7.7) i.e., equations (7.13) and (7.17) determine the *classical effective internuclear potential* for any strength of the magnetic field, including the strong field case.

Figure 7.4 shows the upper and middle branches of the scaled effective internuclear potential v versus the scaled internuclear distance r for $Z = 2$ and $Z' = 4$ in the absence of the magnetic field. It is seen for any starting point at the middle branch, that the system would 'find' the way to lowering its potential energy without any obstacle and would end up at an infinitely large internuclear distance, thereby

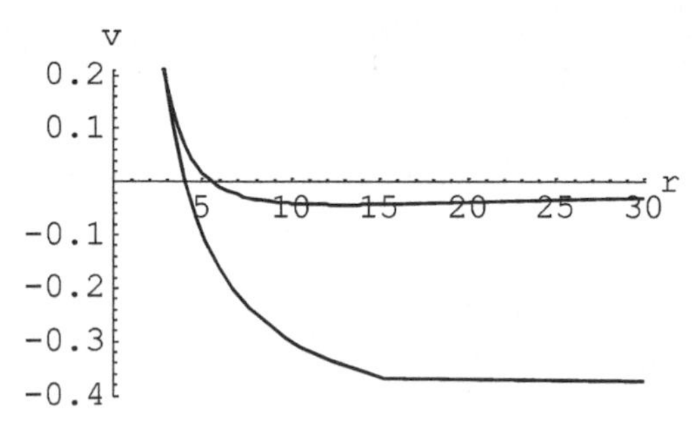

Figure 7.4. The upper and middle branches of the scaled effective internuclear potential v (defined by equations (7.15), (7.16)) versus the scaled internuclear distance r for $Z = 2$ and $Z' = 4$ at the absence of the magnetic field.

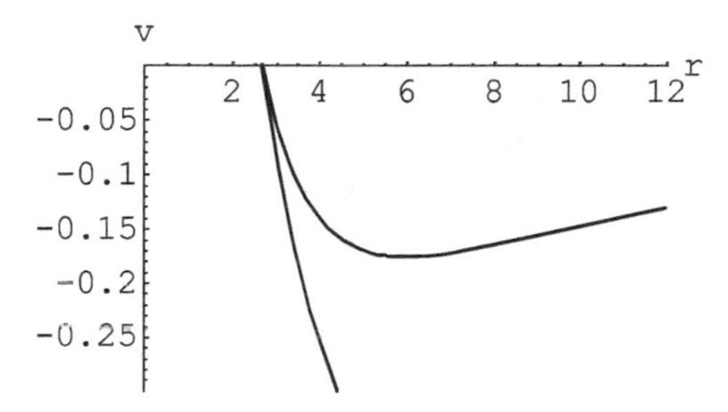

Figure 7.5. The same as in figure 7.4, but at the scaled magnetic field $\omega = -0.3$ (note that $BM < 0$).

resulting in dissociation. The same is true for the lower branch (not shown in figure 7.4). In other words, in the absence of the magnetic field, the CRS-system, associated with the middle or lower branches of the effective potential energy, is not really a molecule, but only a quasimolecule because the molecular orbital is *antibonding*. As we noted, the corresponding classical result was obtained previously by Pauli [10] for the molecular hydrogen ion H_2^+. The upper branch in figure 7.4 displays a very shallow minimum of $v = -10.0444$ located at $r = 13.0$.

We now 'turn on' the magnetic field. Figure 7.5 shows the upper and middle branches of the scaled effective internuclear potential v versus the scaled internuclear distance r for $Z = 2$ and $Z' = 4$ at a relatively small scaled magnetic field $\omega = -0.3$ (with $BM < 0$)[1]. It is seen that the minimum in the upper branch became significantly deeper and moved towards lower r.

Figure 7.6 shows the same as figure 7.5, but for $\omega = -1$. As the magnetic field increased, it is seen that the minimum in the upper branch becomes further deepened and moves even closer to the origin.

The 'cusp' formed by the upper and middle branches in figures 7.4–7.6 reflects the fact that the upper and middle energy terms for the corresponding electronic terms terminate at some $r = r_{min}$—as already noted above. Although present in CRS, this

[1] The scaled magnetic field $|\omega| = 0.3$ would correspond to the magnetic field $B \sim 10^5$ G for $|M| \sim 30$. The magnetic field $B \sim 10^5$ G would be typical for magnetic fusion devices under construction.

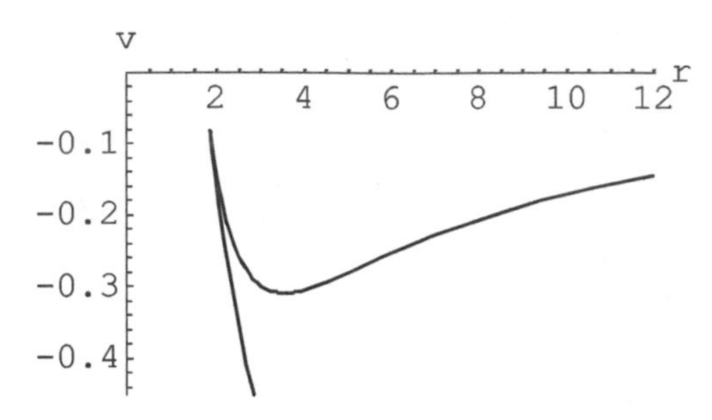

Figure 7.6. The same as in figure 7.4, but at the scaled magnetic field $\omega = -1$ (note that $BM < 0$).

cusp may not appear in non-circular Rydberg states, a topic beyond the scope of paper [1].

Figures 7.4–7.6 reveal *magnetic stabilization of the nuclear motion* for the case of $BM < 0$. Indeed, in the absence of the magnetic field, the potential well is very shallow. It is known that too shallow potential wells do not have any quantal discrete energy levels (see, e.g., the book [6]). Moreover, if this system is embedded in a plasma, then due to the known phenomenon of the 'continuum lowering' by the plasma environment (see, e.g., books/reviews [16–18] and references therein), the minimum of this very shallow potential well in figure 7.4 could be 'absorbed' by the lowered continuum. The magnetic field dramatically deepens the potential well and therefore stabilizes the system for the case of $BM < 0$. The magnetic field can therefore transform the quasimolecule into a real, classically described molecule so that the molecular orbital becomes *bonding*.

The particular example of the system chosen for figures 7.4–7.6 corresponds to the CRS of an electron in the vicinity of the nuclei of He and Be (in paper [1] the examples were for the CRS of an electron in the vicinity of the nuclei of He and Li). All of the above nuclei could be present in magnetic fusion plasmas. Moreover, in such plasmas, Rydberg states of either of these nuclei result from charge exchange with ions of higher nuclear charge that are always present in magnetic fusion plasmas. Relatively large magnetic-field strengths are also present. It should be therefore possible to observe magnetic stabilization of the quasimolecule $HeLi^{4+}$ or $HeBe^{5+}$ present in these practically important experimental devices.

The analysis in paper [1] has also shown that a similar magnetic stabilization of Rydberg quasimolecules in CRS is displayed by other (though not all) ZeZ'-systems characterized by the ratio of the nuclear charges in the range: $1 < Z'/Z < 3$. These results open up this phenomenon for possible further theoretical and experimental investigation.

References

[1] Flannery M R and Oks E 2006 *Phys. Rev.* A **73** 013405

[2] Oks E 2000 *Phys. Rev. Lett.* **85** 2084

[3] Oks E 2000 *J. Phys. B: Atom. Mol. Opt. Phys.* **33** 3319

[4] Pauli W 1922 *Ann. d. Physik* **68** 177

[5] Landau L D and Lifshitz E M 1960 *Mechanics* (Oxford: Pergamon)
[6] Landau L D and Lifshitz E M 1965 *Quantum Mechanics* (Oxford: Pergamon)
[7] Born M and Oppenheimer R 1927 *Ann. Physik* **84** 457
[8] Gershtein S S and Krivchenkov V D 1961 *Sov. Phys. JETP* **13** 1044
[9] Von Neumann J and Wigner E 1929 *Phys. Z.* **30** 467
[10] Ponomarev L I and Puzynina T P 1967 *Sov. Phys. JETP* **25** 846
[11] Power J D 1973 *Phil. Trans. R. Soc. London* **A274** 663
[12] Komarov I V, Ponomarev L I and Slavyanov S Y 1976 *Spheroidal and Coulomb Spheroidal Functions* (Moscow: Nauka) in Russian
[13] Garton W R S and Tomkins F S 1969 *Astrophys. J.* **158** 839
[14] Brack M and Bhaduri R K 1997 *Semiclassical Physics* (Reading, MA: Addison-Wesley) section 1.4.2
[15] Galitski V, Karnakov B, Kogan V and Galitski V Jr. 2013 *Exploring Quantum Mechanics* (Oxford: Oxford University Press) Problem 8.55
[16] Salzmann D 1998 *Atomic Physics in Hot Plasmas* (Oxford: Oxford University Press) ch 2, 3
[17] Murillo M S and Weisheit J C 1998 *Phys. Rep.* **302** 1
[18] Griem H R 1997 *Principles of Plasma Spectroscopy* (Cambridge: Cambridge University Press) sections 5.5, 7.3

IOP Publishing

Analytical Advances in Quantum and Celestial Mechanics

Separating rapid and slow subsystems

Eugene Oks

Chapter 8

One-electron Rydberg quasimolecules in a high-frequency laser field

The problem of electron terms in the field of two stationary Coulomb centers (TCC) of charges Z and Z' separated by a distance R is one of the most fundamental problems in quantum mechanics. It presents fascinating atomic physics: the terms can have crossings and quasicrossings. On the one hand, the well-known Neumann–Wigner general theorem on the impossibility of crossing of terms of the same symmetry [1] is invalidated for the TCC problem of $Z' \neq Z$ (see, e.g., paper [2])—so, the terms can cross. On the other hand, when two potential wells (each corresponding to separated Z- and Z'-centers) have states Ψ and Ψ' of the same energies $E = E'$, of the same magnetic quantum numbers $m = m'$, and of the same radial elliptical quantum numbers $k = k'$, a quasicrossing of the terms occurs [3–5]. Then the electron has a much larger probability of tunneling from one well to the other (what constitutes charge exchange) compared to the absence of the quasicrossing.

In plasma spectroscopy, a quasicrossing of the TCC terms, by facilitating charge exchange, can result in local dips in the spectral line profile emitted by a Z-ion from a plasma consisting of both Z- and Z'-ions—see, e.g., theoretical and experimental papers [6–11]. In particular, this allows determining rates of charge exchange between multicharged ions—the reference data almost inaccessible by other experimental methods [11].

When the charges Z and Z' approach each other and share the only electron that they have, they form a quasimolecule. When the electron is in a highly-excited state, it is a one-electron Rydberg quasimolecule (OERQ). There are extensive analytical studies of the OERQ by the methods of classical mechanics (which are appropriate for Rydberg states) [12–20]—see also review [21] and book [22], chapter 3. In particular, the following papers were devoted to studies of the QERQ in various external fields: namely, in a static magnetic field [15], in a static electric field [16, 17, 19], and in a laser field [20]. Specifically, in our previous paper [20] we analyzed the situation where the laser frequency was much smaller than the highest frequency of the unperturbed system.

In paper [23] the authors considered the situation where the OERQ is subjected to a linearly-polarized laser field whose frequency is much greater than the highest frequency of the unperturbed system. For obtaining analytical results they used a generalization of the method of effective potentials [24], presented also in the current book in appendix A. They showed that as the amplitude of the laser field increases, the structure of the energy terms becomes more and more complicated, and the number of the energy terms increases. They also calculated analytically the shift of the radiation frequency of OERQ caused by the laser field. Here are the details.

The authors of paper [23] considered a TCC system with the charge Z placed at the origin, and the Oz axis is directed at the charge Z', which is at $z = R$. Atomic units ($\hbar = e = m_e = 1$) were used. The system is subjected to a high-frequency linearly-polarized laser field of amplitude F and frequency ω, the laser field being directed along the internuclear axis. The Hamiltonian for the electron in this configuration is

$$H = H_0 + zF\cos\omega t,\ H_0 = \frac{1}{2}\left(p_z^2 + p_\rho^2 + \frac{p_\phi^2}{\rho^2}\right) - \frac{Z}{r} - \frac{Z'}{r'} \tag{8.1}$$

where $r = (\rho^2 + z^2)^{1/2}$ is the distance from the electron to the nucleus Z, $r' = (\rho^2 + (R - z)^2)^{1/2}$ is the distance from the electron to the nucleus Z', and (ρ, φ, z) are the cylindrical coordinates positioned in such a way that the nuclei Z and Z' are on the z-axis at $z = 0$ and $z = R$ accordingly. Due to φ-symmetry, φ is a cyclic coordinate and its corresponding momentum is conserved:

$$p_\phi = \rho^2\frac{d\phi}{dt} = L \tag{8.2}$$

For the systems in a high-frequency field, whose frequency is much greater than the highest frequency of the unperturbed system, it is appropriate to use the formalism of effective potentials [24–26]. As a result, the Hamiltonian acquires a time-independent term. The zeroth-order effective potential,

$$U_0 = \frac{1}{4\omega^2}[V, [V, H_0]] = \frac{F^2}{4\omega^2} \tag{8.3}$$

where $V = zF$ and $[P, Q]$ are the Poisson brackets, is a coordinate-independent energy shift that does not affect the dynamics of the system. The first non-vanishing effect on the dynamics of the system originates from the first-order effective potential

$$\begin{aligned} U_1 &= \frac{1}{4\omega^4}[[V, H_0], [[V, H_0], H_0]] \\ &= \frac{F^2}{4\omega^4}\left(Z\frac{\rho^2 - 2z^2}{(\rho^2 + z^2)^{5/2}} + Z'\frac{\rho^2 - 2(R - z)^2}{(\rho^2 + (R - z)^2)^{5/2}}\right) \end{aligned} \tag{8.4}$$

and the Hamiltonian of the electron in the high-frequency field is

$$H = \frac{1}{2}\left(p_z^2 + p_\rho^2\right) + \frac{L^2}{2\rho^2} - \frac{Z}{\sqrt{\rho^2 + z^2}} - \frac{Z'}{\sqrt{\rho^2 + (R - z)^2}} + U_1 \tag{8.5}$$

where U_1 is given by equation (8.4). The electron is considered to be in a circular state[1]. Therefore, $p_z = p_\rho = 0$, and thus, its energy can be represented in the form

$$\begin{aligned} E = &\frac{L^2}{2\rho^2} - \frac{Z}{\sqrt{\rho^2 + z^2}} - \frac{Z'}{\sqrt{\rho^2 - (R - z)^2}} \\ &+ \frac{F^2}{4\omega^4}\left(Z\frac{\rho^2 - 2z^2}{(\rho^2 + z^2)^{5/2}} + Z'\frac{\rho^2 - 2(R - z)^2}{(\rho^2 + (R - z)^2)^{5/2}}\right) \end{aligned} \tag{8.6}$$

Using the scaled quantities

$$w = \frac{z}{R},\ v = \frac{\rho}{R},\ \varepsilon = -\frac{R}{Z}E,\ b = \frac{Z'}{Z},\ \ell = \frac{L}{\sqrt{ZR}},\ r = \frac{Z}{L^2}R,\ \theta = \frac{F}{\omega^2 R} \tag{8.7}$$

one obtains the scaled energy of the electron

$$\begin{aligned} \varepsilon = &\frac{1}{\sqrt{w^2 + v^2}} + \frac{b}{\sqrt{(1 - w)^2 + v^2}} \\ &- \frac{\ell^2}{2v^2} + \frac{2w^2 - v^2}{(w^2 + v^2)^{5/2}}\frac{\theta^2}{4} + b\frac{2(1 - w)^2 - v^2}{((1 - w)^2 + v^2)^{5/2}}\frac{\theta^2}{4} \end{aligned} \tag{8.8}$$

One can seek the equilibrium points in the (w, v)-plane by finding the two partial derivatives of ε with respect to w and v and setting them to zero. The second equation gives the equilibrium value of the scaled angular momentum

$$\ell = v^2\sqrt{\frac{(w^2 + v^2)^2 + 3\left(w^2 - \frac{v^2}{4}\right)\theta^2}{(w^2 + v^2)^{7/2}} + b\frac{((1 - w)^2 + v^2)^2 + 3\left((1 - w)^2 - \frac{v^2}{4}\right)\theta^2}{((1 - w)^2 + v^2)^{7/2}}} \tag{8.9}$$

and the first equation gives the equilibrium value of v

[1] Circular states of atomic and molecular systems are an important subject. They have been extensively studied both theoretically and experimentally for several reasons (see, e.g., [12–15, 17, 27–40] and references therein): (a) they have long radiative lifetimes and highly anisotropic collision cross sections, thereby enabling experiments on inhibited spontaneous emission and cold Rydberg gases, (b) these classical states correspond to quantal coherent states, objects of fundamental importance, (c) a classical description of these states is the primary term in the quantal method based on the $1/n$-expansion, and (d) they can be used in developing atom chips.

$$w\frac{(w^2+v^2)^2+\frac{3}{2}\left(w^2-\frac{3}{2}v^2\right)\theta^2}{(w^2+v^2)^{7/2}}$$
$$=b(1-w)\frac{((1-w)^2+v^2)^2+\frac{3}{2}\left((1-w)^2-\frac{3}{2}v^2\right)\theta^2}{((1-w)^2+v^2)^{7/2}} \tag{8.10}$$

Figure 8.1 shows the equilibrium plot in the (w, v)-plane for $b = 3$ and $\theta = 0.1$. It is seen that, in addition to the properties described in paper [13], there is a multivalued range in the neighborhood of $w = 0$ and $w = 1$, which increases as θ increases.

If one scales the internuclear distance R as $r = (Z/L^2)R$, and given $\varepsilon = -(R/Z)E$ from equation (8.7), then $E = -(Z/L)^2\,\varepsilon_1$, where $\varepsilon_1 = \varepsilon/r$ is the scaled energy whose scaling includes only Z and L. From equation (8.7), $\ell^2 = L^2/(ZR)$, so this yields $r = 1/\ell^2$, with ℓ taken from equation (8.9), giving us the expression for $r(w, v, b, \theta)$. Then by substituting the value of ℓ from equation (8.9) into equation (8.8) and obtaining $\varepsilon(w, v, b, \theta)$, subsequently divided by $r = 1/\ell^2$, with ℓ again taken from equation (8.9), one obtains $\varepsilon_1(w, v, b, \theta)$, whose explicit form is

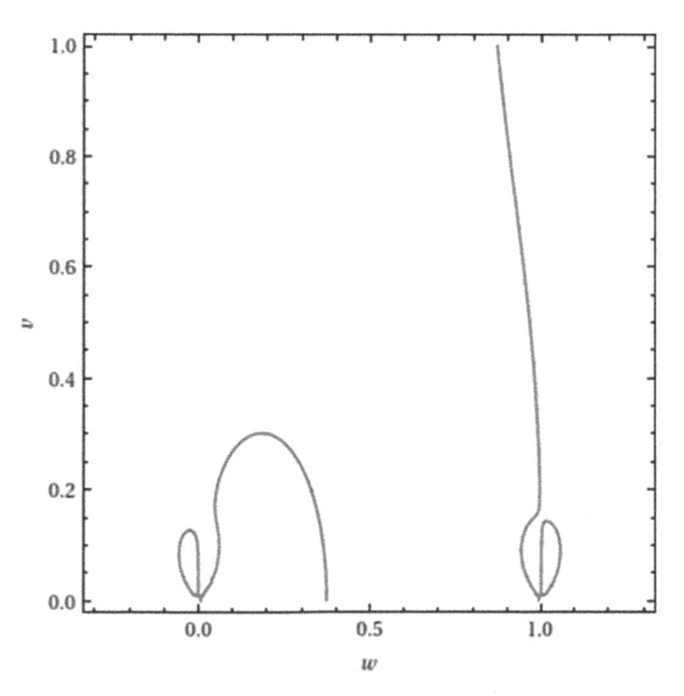

Figure 8.1. Equilibrium plot in the (w, v)-plane for $b = 3$ and $\theta = 0.1$. (Reproduced with permission from [23]. Copyright 2019 N Kryukov and E Oks.)

$$\varepsilon_1 = v^4 \left(\frac{(w^2+v^2)^2\left(w^2+\frac{v^2}{2}\right)+\left(w^4-\frac{5}{2}w^2v^2+\frac{v^4}{4}\right)\frac{\theta^2}{2}}{(w^2+v^2)^{7/2}} \right.$$
$$\left. +b\frac{((1-w)^2+v^2)^2\left((1-w)^2+\frac{v^2}{2}\right)+\left((1-w)^4-\frac{5}{2}(1-w)^2v^2+\frac{v^4}{4}\right)\frac{\theta^2}{2}}{((1-w)^2+v^2)^{7/2}} \right) \tag{8.11}$$
$$\times \left(\frac{(w^2+v^2)^2+3\left(w^2-\frac{v^2}{4}\right)\theta^2}{(w^2+v^2)^{7/2}} + b\frac{((1-w)^2+v^2)^2+3\left((1-w)^2-\frac{v^2}{4}\right)\theta^2}{((1-w)^2+v^2)^{7/2}} \right)$$

Then, by solving equation (8.10) numerically for v and substituting it into equation (8.11) and into $r(w, v, b, \theta)$, one obtains, for the given value of b and θ, the parametric dependence $\varepsilon_1(r)$ representing the scaled energy terms, with the parameter w running over the allowed range determined by equation (8.10). The asymptote w_3, corresponding to $v \to \infty$, is the same as in the case of $\theta = 0$, and is equal to $b/(b + 1)$, and other limits on w can be determined numerically.

Figures 8.2 and 8.3 show the scaled energy terms for the values of the scaled amplitude of the laser field for $\theta = 0.01$ and $\theta = 0.1$, respectively, in comparison to the unperturbed energy terms for $\theta = 0$. It is seen that for small values of θ the lower term is the first affected, and the terms take on a more complicated form as θ further increases. The quantity $-\varepsilon_1$ is plotted on the vertical axis for it to have the same sign as E.

It is seen that as the scaled amplitude θ of the laser field increases, the scaled energy terms $-\varepsilon_1(r)$ become more and more complicated. In particular, at some ranges of θ, the number of the scaled energy terms increases from 3 (which was the case for $\theta = 0$) to 4 or even 5.

At this point it might be useful to clarify the relation between the classical energy terms $-\varepsilon_1(r)$ and the energy E. The former is a scaled quantity related to the energy as specified above in the 1st line after equation (8.10): $E = -(Z/L)^2\,\varepsilon_1$. The projection L of the angular momentum on the internuclear axis is a *continuous* variable. The energy E depends on both ε_1 and L. Therefore, while the scaled quantity ε_1 takes a *discrete* set of values, the energy E takes a *continuous* set of values (as it should be in classical physics).

The authors of paper [23] also studied the shift of the radiation frequency caused by a high-frequency linearly-polarized laser field. The angular momentum of the electron can be expressed as

$$L = \rho^2\frac{\mathrm{d}\phi}{\mathrm{d}t} = \Omega\rho^2 \tag{8.12}$$

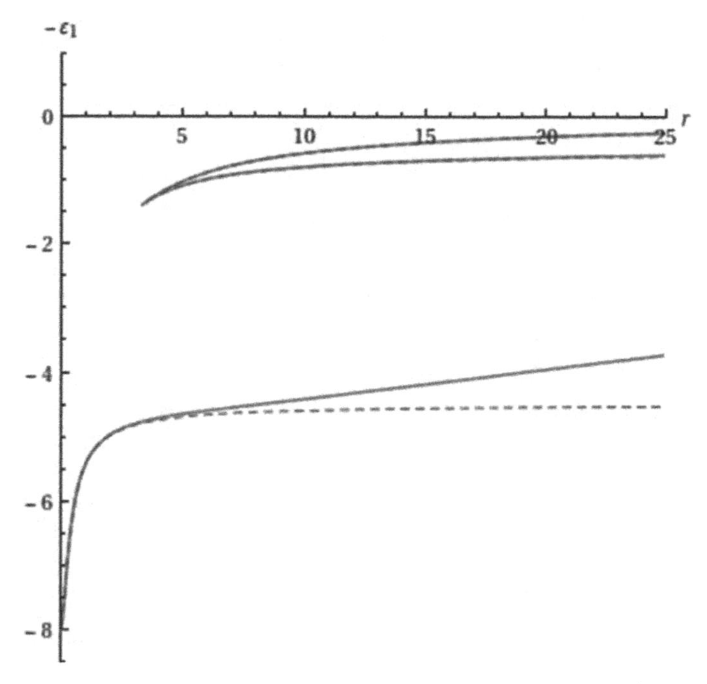

Figure 8.2. The plot of the scaled energy terms $-\varepsilon_1(r)$ (with r on the horizontal axis and $-\varepsilon_1$ on the vertical) for the scaled amplitude of the laser field $\theta = 0.01$, with $b = 3$, shown in blue, solid curves, against the terms for $\theta = 0$, with $b = 3$, shown in red, dashed curves. (Reproduced with permission from [23]. Copyright 2019 N Kryukov and E Oks.)

where Ω is the frequency of the motion of the electron. Using the scaled quantities from equation (8.7), one has

$$\Omega = \sqrt{\frac{Z}{R^3}}\tilde{\Omega},\ \tilde{\Omega} = \frac{\ell}{v^2} \tag{8.13}$$

where Ω with the wave above denotes the scaled frequency. The relative shift of the frequency is determined by

$$\delta = \frac{\Omega - \Omega_0}{\Omega_0} = \frac{\Omega}{\Omega_0} - 1 = \frac{\tilde{\Omega}}{\tilde{\Omega}_0} - 1 = \frac{\ell}{\ell_0}\frac{v_0^2}{v^2} - 1 \tag{8.14}$$

where the subscript index '0' refers to the default case ($\theta = 0$) and the value of v is taken to be the equilibrium value (determined by equation (8.10)).

Figures 8.4 and 8.5 show the plot of the relative shift of the frequency for the ratio of the nuclear charges $b = 3$ and the values of $\theta = 0.01$ and $\theta = 0.1$, respectively. It is seen, the shift increases when θ increases, and it is the smallest around the point $w = w_3 = b/(b + 1)$.

Thus, for a known amplitude of the laser field, by measuring the relative shift of the radiation frequency it should be possible to determine experimentally the

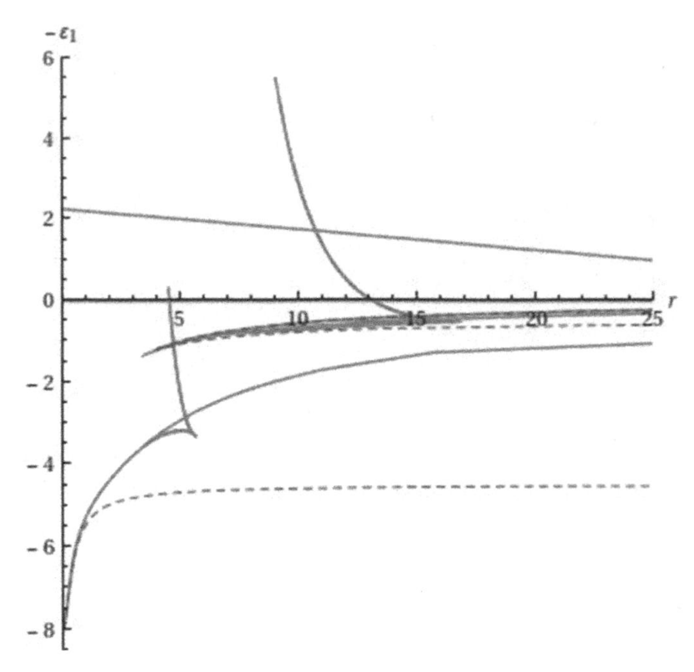

Figure 8.3. The plot of the scaled energy terms $-\varepsilon_1(r)$ (with r on the horizontal axis and $-\varepsilon_1$ on the vertical) for the scaled amplitude of the laser field $\theta = 0.1$, with $b = 3$, shown in blue, solid curves, against the terms for $\theta = 0$, with $b = 3$, shown in red, dashed curves. (Reproduced with permission from [23]. Copyright 2019 N Kryukov and E Oks.)

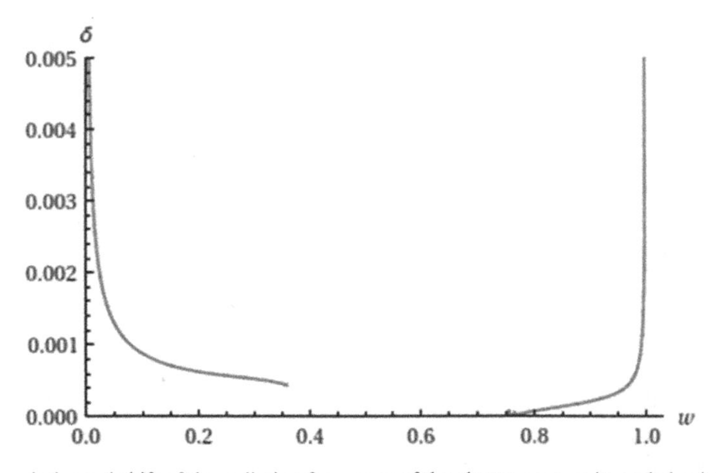

Figure 8.4. The relative red shift of the radiation frequency of the electron versus its scaled axial coordinate for $b = 3$ and $\theta = 0.01$. (Reproduced with permission from [23]. Copyright 2019 N Kryukov and E Oks.)

distance of the orbital plane of the electron from the nucleus of the smaller nuclear charge.

In summary, in paper [23] the authors considered the situation where one-electron Rydberg quasimolecules (OERQ) are subjected to a linearly-polarized laser field

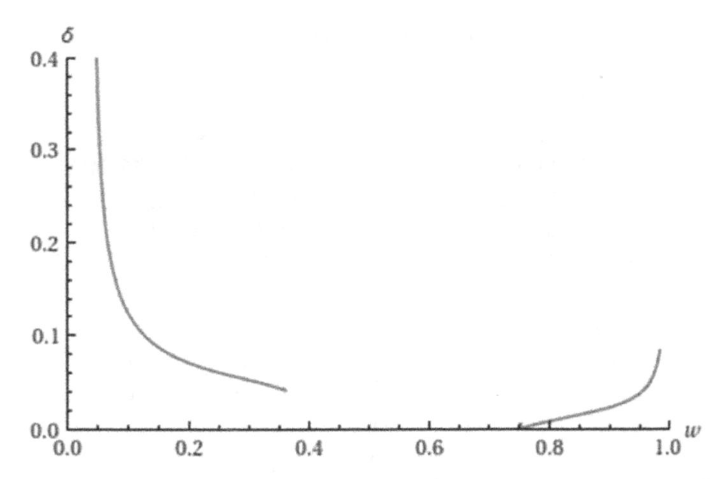

Figure 8.5. The relative red shift of the radiation frequency of the electron versus its scaled axial coordinate for $b = 3$ and $\theta = 0.1$. (Reproduced with permission from [23]. Copyright 2019 N Kryukov and E Oks.)

whose frequency is much greater than the highest frequency of the unperturbed system. They found out that as the amplitude of the laser field increases, the structure of the energy terms becomes more and more complicated. Moreover, the number of the energy terms increases.

They also calculated analytically the shift of the radiation frequency of OERQ caused by the laser field. As the amplitude of the laser field increases, so does the shift. For a known amplitude of the laser field, by measuring the relative shift of the radiation frequency it should be possible to determine experimentally the distance of the orbital plane of the electron from the nucleus of the smaller nuclear charge.

References

[1] Von Neumann J and Wigner E 1929 *Phys. Z.* **30** 467
[2] Gershtein S S and Krivchenkov V D 1961 *Sov. Phys. JETP* **13** 1044
[3] Ponomarev L I and Puzynina T P 1967 *Sov. Phys. JETP* **25** 846
[4] Power J D 1973 *Phil. Trans. R. Soc. Lond.* **A274** 663
[5] Komarov I V, Ponomarev L I and Slavyanov S Y 1976 *Spheroidal and Coulomb Spheroidal Functions* (Moscow: Nauka) in Russian
[6] Böddeker S, Kunze H-J and Oks E 1995 *Phys. Rev. Lett.* **75** 4740
[7] Oks E and Leboucher-Dalimier E 2000 *Phys. Rev. E, Rap. Comm.* **62** R3067
[8] Oks E and Leboucher-Dalimier E 2000 *J. Phys.* B **33** 3795
[9] Leboucher-Dalimier E, Oks E, Dufour E, Sauvan P, Angelo P, Schott R and Poquerusse A 2001 *Phys. Rev. E, Rap. Comm.* **64** 065401
[10] Leboucher-Dalimier E, Oks E, Dufour E, Angelo P, Sauvan P, Schott R and Poquerusse A 2002 *Eur. Phys. J.* D **20** 269
[11] Dalimier E, Oks E, Renner O and Schott R 2007 *J. Phys.* B **40** 909
[12] Oks E 2000 *Phys. Rev. Lett.* **85** 2084
[13] Oks E 2000 *J. Phys. B: Atom. Mol. Opt. Phys.* **33** 3319
[14] Oks E 2001 *Phys. Rev.* E **63** 057401

[15] Flannery M R and Oks E 2006 *Phys. Rev.* A **73** 013405
[16] Kryukov N and Oks E 2011 *Inter. Rev. Atom. Mol. Phys.* **2** 57
[17] Kryukov N and Oks E 2012 *Canad. J. Phys.* **90** 647
[18] Kryukov N and Oks E 2012 *Inter. Rev. Atom.Mol. Phys.* **3** 113
[19] Kryukov N and Oks E 2013 *J. Phys. B: Atom. Mol. Opt. Phys.* **46** 245701
[20] Kryukov N and Oks E 2014 *Europ. Phys. J.* D **68** 171
[21] Kryukov N and Oks E 2013 *Inter. Rev. Atom. Mol. Phys.* **4** 121
[22] Oks E 2015 *Breaking Paradigms in Atomic and Molecular Physics* (Singapore: World Scientific)
[23] Kryukov N and Oks E 2019 *Inter. Rev. Atom. and Mol. Phys.* **9** 1
[24] Nadezhdin B B 1986 *Radiatsionnye i Relativistskie Effekty v Atomakh i Ionakh (Radiative and Relativistic Effects in Atoms and Ions* (Moscow: Scientific Council of the USSR Academy of Sciences on Spectroscopy), 222 in Russian
[25] Kapitza P L 1951 *Sov. Phys. JETP* **21** 588
[26] Kapitza P L 1951 *Uspekhi Fiz. Nauk* **44** 7
[27] Mishra A P, Nandi T and Jagatap B N 2013 *J. Quant. Spectrosc. Rad. Transf.* **120** 114
[28] Flannery M R and Oks E 2008 *Eur. Phys. J.* D **47** 27
[29] Nogues G, Lupascu A, Emmert A, Brune M, Raimond J-M and Haroche S 2011 *Atom Chips* ed J Reichel and V Vuletic (Weinheim, Germany: Wiley-VCH) ch 10, section 10.3.3
[30] Tan J N, Brewer S M and Guise N D 2011 *Phys. Scripta* **T144** 014009
[31] Kryukov N and Oks E 2012 *Intern. Rev. Atom. Mol. Phys.* **3** 17
[32] Dehesa J S, Lopez-Rosa S, Martinez-Finkelshtein A and Janez R J 2010 *Intern. J. Quant. Chem.* **110** 1529
[33] Nandi T 2009 *J. Phys. B: At. Mol. Opt. Phys.* **42** 125201
[34] Jentschura U D, Mohr P J, Tan J N and Wundt B J 2008 *Phys. Rev. Lett.* **100** 160404
[35] Shytov A V, Katsnelson M I and Levitov L S 2007 *Phys. Rev. Lett.* **99** 246802
[36] Devoret M, Girvin S and Schoelkopf R 2007 *Ann. Phys.* **16** 767
[37] Oks E 2004 *Eur. Phys. J.* D **28** 171
[38] Holmlid L 2002 *J. Phys.: Cond. Matt.* **14** 13469
[39] Dutta S K, Feldbaum D, Walz-Flannigan A, Guest J R and Raithel G 2001 *Phys. Rev. Lett.* **86** 3993
[40] Carlsen H and Goscinski O 1999 *Phys. Rev.* A **59** 1063

IOP Publishing

Analytical Advances in Quantum and Celestial Mechanics
Separating rapid and slow subsystems
Eugene Oks

Chapter 9

Quantum rotator-dipole in a high-frequency monochromatic field: a violation of the gauge invariance caused by the constraint

In this chapter we analyze first a shift of rotational energy levels of a rigid rotator, having an electric dipole moment, under a high-frequency monochromatic linearly-polarized electric field. We note that Braun and Petelin [1] treated this problem by using Kapitza's effective potential [2–4]. The Hamiltonian of a free rigid rotator has the form

$$H_r = -B[(1/\sin\theta)\partial/\partial\theta\,(\sin\theta\partial/\partial\theta) + (1/\sin^2\theta)\partial^2/\partial\varphi^2, \tag{9.1}$$

where B is the rotational constant; θ and φ are the polar and azimuthal angles of the rotator, respectively. Following Nadezhdin [5], for simplicity of formulas we assume that the rotator is formed by two point-like particles of the masses m and of the charges $+e$ and $-e$, separated by a distance d. Then $B = \hbar^2/(md^2)$ and the dipole moment of the rotator is equal to ed.

The interaction of the rotator with a linearly-polarized electric field $\mathbf{F}\cos\omega t$ can be represented in the following two gauges V or W (the z-axis being chosen parallel to vector $\mathbf{F}$):

$$\begin{aligned} V &= -(edF\cos\theta)\cos\omega t,\\ W &= -\sin\omega t[2ie\hbar A/(mcd)](\sin\theta)\partial/\partial\theta, \quad A = cF/\omega. \end{aligned} \tag{9.2}$$

In gauge W, the coordinate-independent term containing A^2 has been omitted. The field is considered to be the high-frequency one, meaning that

$$\hbar\omega \gg B. \tag{9.3}$$

The Kapitza's effective potential in V and W gauges, respectively is

$$V_0 = [(edF)^2 B/(2\hbar^2\omega^2)]\sin^2\theta,$$
$$W_0 = -[(eA)^2 B/(\omega mcd)^2](4 - 2\sin^2\theta) \times [(1/\sin\theta)\partial/\partial\theta + (1/\sin^2\theta)\partial^2/\partial\varphi^2]. \quad (9.4)$$

(For details on the formalism of the Kapitza's effective potential see appendix A.) We note that the expression for V_0 in equation (9.4) had been obtained in paper [1].

Obviously, operators V_0 and W_0 are diagonal with respect to the projection M of the angular momentum J on the axis of rotation. The average values of V_0 and W_0 with respect to the unperturbed states $|JM\rangle$ of the rigid rotator are as follows:

$$\langle JM|V_0|JM\rangle = [(eF)^2/(m\omega^2)](J^2 + J - 1 + M^2)/[(2J - 1)(2J + 3)],$$
$$\langle JM|W_0|JM\rangle = [(eF)^2/(m\omega)^2][\hbar/(m\omega d^2)]^2\, 4J(J + 1)(3J^2 + 3J - 2 - M^2) /[(2J - 1)(2J + 3)]. \quad (9.5)$$

It is seen that the energy shift $E_k^{(2)}$, calculated with the help of the Kapitza's effective potential in different gauges, yields different results. In other words, the energy shift $E_k^{(2)}$ turns out to be not gauge invariant. In particular, in the gauge V the energy shift is proportional to $1/\omega^2$, while in the gauge W the energy shift is proportional to $1/\omega^4$. Moreover, it is seen from equation (9.5) that in the two different gauges, the energy shift has different orders of magnitude since $\hbar/(m\omega d^2) \ll 1$.

So, a question arises as to which of the two expressions in equation (9.5) corresponds to the reality. In other words, which of the two gauges yields the correct description of the interaction of the rigid rotator with the high-frequency monochromatic field. Below, following Nadezhdin [5], it is shown that the *violation of the gauge invariance* is caused by the fact that in the rigid rotator the distance d between the two point-like masses is fixed, i.e., a *constraint* is imposed on the motion of these two masses.

Let us consider an oscillating rotator, i.e., the rotator where the distance x between the point-like masses can oscillate around the equilibrium distance d. Such a system can be described by the following Hamiltonian:

$$H = U(x) + \sum_{i=1,2}\sum_{j=1,2,3}[\hbar^2/(2m)]\partial^2/[\partial x_j^{(i)}]^2,$$
$$x = \left[\sum_{j=1,2,3}(x_j^{(1)} - x_j^{(2)})^2\right]^{1/2}. \quad (9.6)$$

Here suffix i is the label of the two masses 1 and 2.

The internal motion of the oscillating rotator can be described in terms of the angular variables θ and φ, determining the spatial orientation of the rotator, and the variable distance x. As usual, we assume the two masses for a linear oscillator, so that

$$U(x) = K(x - d)^2/2, \quad K = \text{const}. \tag{9.7}$$

For sufficiently large K, it is possible to disregard the dependence of the rotator moment of inertia on x (i.e., to consider $|x - d| \ll d$), so that the oscillation and the rotation can be separated, leading to the Hamiltonian

$$H = H_v + H_r, \quad H_v = K(x - d)^2/2 + (\hbar^2/m)\partial^2/\partial x^2, \tag{9.8}$$

where the partial Hamiltonian H_r (the Hamiltonian of the rigid rotator) is given by equation (9.1). In equation (9.8) it was taken into account that the reduced mass of the oscillator is $m/2$. Obviously, the system described by the Hamiltonian H from equation (9.8) does not have constraints.

Eigenfunctions of the Hamiltonian are $|JM\rangle|v\rangle$, where $|JM\rangle$ are the eigenfunctions of the rigid rotator (depending on θ and φ) and $|v\rangle$ are the eigenfunctions of the linear oscillator (depending on x). The unperturbed energy of the oscillating rotator is

$$E^{(0)} = BJ(J + 1) + \hbar\Omega(v + 1/2), \quad \Omega = (2K/m)^{1/2}, \tag{9.9}$$

Ω being the eigenfrequency of the oscillator. Since we are interested in the limiting transition to the rigid rotator, i.e., in the case of $\Omega \to \infty$, then we assume that

$$\hbar\Omega \gg \hbar\omega \gg B. \tag{9.10}$$

The interaction of the oscillating rotator with the electric field can be represented in the following two different gauges (in analogy with equation (9.2)):

$$\begin{aligned} V &= -(eFx\cos\theta)\cos\omega t, \\ W &= \sin\omega t[2ie\hbar A/(mc)][\cos\theta\, \partial/\partial x - (1/d)\sin\theta\, \partial/\partial\theta]. \end{aligned} \tag{9.11}$$

It is seen that equation (9.11) differs from equation (9.2). It is important that while in gauge V the difference is simply the substitution of d by x, in gauge W the difference is much more substantial: in the expression for W in equation (9.11) there is a term containing ∂/∂x, which was absent in the expression for W in equation (9.2). Below it is shown that it is this term that is responsible for the violation of the gauge invariance.

Indeed, let us consider the energy shift

$$E_k^{(2)} = [1/(2\hbar)]\sum_i V_{ki}V_{ik}\omega_{ik}/\left(\omega^2 - \omega_{ik}^2\right), \tag{9.12}$$

(where $\omega_{ik} = (E_i^{(0)} - E_k^{(0)})/\hbar$) and the similar expression with the substitution of V by W, for the ground state of the oscillating rotator: $J = M = v = 0$. The corresponding matrix elements of the operator V and W between the states of the harmonic linear oscillator differ from zero only for the following cases:

$$\langle 0|x|0\rangle = d, \quad \langle 0|x|1\rangle = [\hbar/(m\Omega)]^{1/2}, \quad \langle 0|\partial/\partial x|1\rangle = (m\Omega/\hbar)^{1/2}/2. \tag{9.13}$$

Therefore, in each gauge the sum $E_0^{(2)}$ can be broken down in two sums: the first one over the rotational states of the state $v = 0$ of the oscillator, the second one over the rotational states of the state $v = 1$ of the oscillator, as follows.

In gauge V:

$$\begin{aligned} E_0^{(2)} = &\sum_{J,M} [eFd|\langle 00|\cos\theta|JM\rangle|]^2\, \omega_{JM} / \left[2\hbar\left(\omega^2 - \omega_{JM}^2\right)\right] \\ &+ \sum_{J,M} [eF|\langle 00|\cos\theta|JM\rangle|]^2 (\Omega + \omega_{JM}) / \{2m\Omega[\omega^2 - (\Omega + \omega_{JM})^2]\}. \end{aligned} \tag{9.14}$$

In gauge W:

$$\begin{aligned} E_0^{(2)} = &\sum_{J,M} 2[e\hbar A|\langle 00|(\sin\theta)\, \partial/\partial\theta|JM\rangle|]^2\, \omega_{JM} / [(mcd)^2(\omega^2 - \omega_{JM}^2)] \\ &+ \sum_{J,M} [eA\Omega|\langle 00|\cos\theta|JM\rangle|]^2 (\Omega + \omega_{JM}) / \{2mc^2[\omega^2 - (\Omega + \omega_{JM})^2]\}. \end{aligned} \tag{9.15}$$

For a *rigid* rotator, *only the first sum* in equations (9.14), (9.15) is used because the rigid rotator does not have states of the vibrational quantum number $v = 1$. Therefore, the correct gauge is the one where the second sum goes to zero at $\Omega \to \infty$. From equations (9.14), (9.15) it is seen that at $\Omega \to \infty$, in gauge V the second sum goes to zero, while in gauge W the second sum goes to the following non-zero limit

$$-[e^2A^2/(2mc^2)] \sum_{J,M} |\langle 00|\cos\theta\,|JM\rangle|^2 = -[e^2F^2/(2m\omega^2)]\langle 00|\cos^2\theta|00\rangle. \tag{9.16}$$

Therefore, the interaction of a rigid rotator with the electric field **F** cos ωt *should be represented in gauge V*, but should not be represented in gauge W.

In summary, the above results can be generalized as follows. While representing the interaction of the field **F** cos ωt with a constraints-having-system in different gauges, one could get different results of the energy shift $E_k^{(2)}$. In order to determine the correct gauge, one should consider a limiting transition from some non-constrained system to the constrained system. In particular, if the constrained system is obtained by imposing *rigid* constraints (what corresponds in quantum mechanics to moving a part of the energy spectrum to infinity), then V would be the correct gauge.

Finally, we note that in book [6] in its section 1.3, while considering multiphoton transitions in some 'model' systems having a finite number of levels, the authors recommended using gauge V, which is consistent with the above results. But for frequencies $\omega \gg \omega_{mn}$ (ω_{mn} being the unperturbed separation in the frequency scale between the levels under consideration), the authors of book [6] recommended using gauge W: their justification was that $A \sim cE/\omega$, so that at relatively large ω the vector potential A becomes relatively small. However, the above results of chapter 8 demonstrated that at least for some physical systems one should use gauge V also for frequencies $\omega \gg \omega_{mn}$.

References

[1] Braun P A and Petelin A N 1974 *Sov. Phys. JETP* **39** 775
[2] Kapitza P L 1951 *Sov. Phys. JETP* **21** 588
[3] Kapitza P L 1951 *Uspekhi Fiz. Nauk* **44** 7
[4] Landau L D and Lifshitz E M 1976 *Mechanics* (Amsterdam: Elsevier) section 30
[5] Nadezhdin B B 1986 *Radiatsionnye i Relativistskie Effekty v Atomakh i Ionakh (Radiative and Relativistic Effects in Atoms and Ions)* (Moscow: Scientific Council of the USSR Academy of Sciences on Spectroscopy), p 222 in Russian
[6] Delone N B and Krainov V P 2012 *Atom in Strong Light Fields* (Berlin: Springer)

Chapter 10

Center-of-mass effects for hydrogen atoms in a non-uniform electric field: applications to magnetic fusion, radiofrequency discharges, and flare stars

There is a lot of literature on center-of-mass (CM) effects for hydrogenic atoms/ions in a uniform magnetic field—see, e.g., papers [1–3] and references therein. The CM motion and the relative (internal) motion are coupled in a magnetic field and, rigorously speaking, cannot be separated. For hydrogen atoms it is possible to achieve a pseudoseparation leading to a Hamiltonian for the relative motion that still depends on a CM integral of the motion called pseudomomentum [3].

As for hydrogenic atoms/ions in a uniform electric field, it is well-known that the CM and relative motions can be separated rigorously (exactly)—see, e.g., [4]. As for hydrogenic atoms/ions in a nonuniform electric field, there seemed to be nothing about the separation (or non-separation) of the CM and relative motions in the literature—until papers [5, 6] were published.

In papers [5, 6] the author studied this issue for hydrogenic atoms/ions in a nonuniform electric field and obtained the following results. First, it was shown that in the general problem of two charges in a nonuniform electric field, the CM and relative motions, rigorously speaking, cannot be separated. Second, there was used an approximate analytical method of the separation of rapid and slow subsystems to achieve a pseudoseparation of the CM and relative motions for hydrogenic atoms/ions in an arbitrary nonuniform electric field. Third, these results were further developed for the case of a hydrogen atom in the nonuniform electric field, where the field is due to the nearest (to the hydrogen atom) ion in a plasma. Fourth, the results were applied to the ion dynamical Stark broadening of hydrogen lines in plasmas. Fifth, there were present specific examples of laboratory and astrophysical plasmas

doi:10.1088/2053-2563/ab3db0ch10

where the allowance for these CM effects leads to a significant increase of the width of hydrogen spectral lines. Here we present details following papers [5, 6].

We consider a system of two charges e_1 and e_2 of masses m_1 and m_2, respectively, in a nonuniform electric field. The Lagrangian of the system is

$$L = [m_1(d\mathbf{r}_1/dt)^2 + m_2(d\mathbf{r}_2/dt)^2]/2 - e_1e_2/|\mathbf{r}_2 - \mathbf{r}_1| - e_1\varphi(\mathbf{r}_1) - e_2\varphi(\mathbf{r}_2), \quad (10.1)$$

where $\mathbf{r}_1$ and $\mathbf{r}_2$ are radii-vectors of charges e_1 and e_2, respectively, and φ is the potential of the nonuniform electric field. After the substitution

$$\mathbf{R} = (m_1\mathbf{r}_1 + m_2\mathbf{r}_2)/(m_1 + m_2), \quad \mathbf{r} = \mathbf{r}_2 - \mathbf{r}_1, \quad (10.2)$$

so that $\mathbf{R}$ and $\mathbf{r}$ are the coordinates related to the CM motion and the relative motion, respectively, the Lagrangian takes the form

$$L(\mathbf{R}, \mathbf{r}) = L_{CM}(\mathbf{R}) - U(\mathbf{R}, \mathbf{r}) + L_r(\mathbf{r}), \quad (10.3)$$

where

$$L_{CM}(\mathbf{R}) = (m_1 + m_2)(d\mathbf{R}/dt)^2/2 - (e_1 + e_2)\varphi(\mathbf{R}) \quad (10.4)$$

is the Langrangian of the CM,

$$L_r(\mathbf{r}) = \mu(d\mathbf{r}/dt)^2/2 - e_1e_2/r \quad (10.5)$$

is the Lagrangian of the relative motion, and

$$U(\mathbf{R}, \mathbf{r}) = \mu(e_1/m_1 - e_2/m_2)\mathbf{r}\mathbf{F}(\mathbf{R}) \quad (10.6)$$

is the coupling of the CM and relative motions. Here

$$\mu = m_1m_2/(m_1 + m_2) \quad (10.7)$$

is the reduced mass of the two particles, and

$$\mathbf{F}(\mathbf{R}) = -d\varphi(\mathbf{R})/d\mathbf{R} \quad (10.8)$$

is a nonuniform electric field (in the expansion of the electric potential we disregarded terms higher than the dipole one). In equation (10.6) and below, for any two vectors **A** and **B**, the notation **AB** stands for the scalar product (also known as the dot-product) of the two vectors.

The Hamiltonian, corresponding to the Langrangian from equation (10.3), has the form

$$H = H_{CM}(\mathbf{R}, \mathbf{P}) + U(\mathbf{R}, \mathbf{r}) + H_r(\mathbf{r}, \mathbf{p}), \quad (10.9)$$

where

$$H_{CM}(\mathbf{R}, \mathbf{P}) = P^2/[2(m_1 + m_2)] + (e_1 + e_2)\varphi(\mathbf{R}) \quad (10.10)$$

is the Hamiltonian of the CM, **P** being the momentum of the CM motion, and

$$H_r(\mathbf{r}, \mathbf{p}) = p^2/(2\mu) + e_1e_2/r \quad (10.11)$$

is the Hamiltonian of the relative motion, **p** being the momentum of the relative motion.

Thus, the above equations show that at the presence of a nonuniform electric field, the CM motion and the relative motion are coupled (by $U(\mathbf{R}, \mathbf{r})$ from equation (10.6)) and therefore, rigorously speaking, cannot be separated. However, in the case where $m_1 \ll m_2$, the CM and relative motions can be separated by using the approximate analytical method of separating rapid and slow subsystems: in this case, the characteristic frequency of the relative motion is much greater than the characteristic frequency of the CM motion, so that the former and the latter are the rapid and slow subsystems, respectively. Below are the details of this method that can be found, e.g., in [7].

The first step is to freeze the coordinates **R** of the slow subsystem and to solve for the motion of the rapid subsystem characterized by the truncated Hamiltonian

$$H_{\rm tr} = H_r(\mathbf{r}, \mathbf{p}) + U(\mathbf{R}, \mathbf{r}) = p^2/(2\mu) + e_1e_2/r + \mu(e_1/m_1 - e_2/m_2)\mathbf{r}\mathbf{F}(\mathbf{R}), \quad (10.12)$$

where **R** is treated as a fixed parameter rather than as the dynamical variable. In the situation where the charges e_1 and e_2 are of the opposite sign (say, for definiteness $e_1 < 0$ and $e_2 > 0$), this becomes the Hamiltonian of a hydrogenic atom/ion in a 'uniform' electric field.

By treating the last term in equation (10.12) in the first order of the perturbation theory, one obtains the following expression for the energy of the relative motion, i.e., the rapid subsystem (see, e.g., the textbook [8])

$$\begin{aligned} E(\mathbf{R}) &= -\mu e_1^2e_2^2/(2n^2\hbar^2) + \mu(e_1/m_1 - e_2/m_2)\langle\mathbf{r}\rangle\mathbf{F}(\mathbf{R}) \\ &= -\mu e_1^2e_2^2/(2n^2\hbar^2) - (3n^2\hbar^2/2)[1/(m_1e_2) + 1/(m_2|e_1|)]\mathbf{A}\mathbf{F}(\mathbf{R}), \end{aligned} \quad (10.13)$$

where there was used the well-known relation between the mean value $\langle\mathbf{r}\rangle$ of the radius-vector and the Runge–Lenz vector **A** (see, e.g., [9, 10]):

$$\langle r\rangle = -3e_1e_2A/|E_0|, \quad E_0 = -\mu e_1^2e_2^2/(2n^2\hbar^2). \quad (10.14)$$

Here and below n is the principal quantum number.

By choosing the z-axis along the Runge–Lenz vector **A**, we rewrite equation (10.13) in the form

$$\begin{aligned} E(\mathbf{R}) = &-\mu e_1^2e_2^2/(2n^2\hbar^2) \\ &- (3n|q|\hbar^2/2)[1/(m_1e_2) + 1/(m_2|e_1|)]F(\mathbf{R})\cos[\theta(\mathbf{R})], \end{aligned} \quad (10.15)$$

where $\theta(\mathbf{R})$ is the polar angle of the vector $\mathbf{F}(\mathbf{R})$ and q is the electric quantum number ($q = n_1 - n_2$, where n_1 and n_2 are the parabolic quantum numbers). Physically, the quantum number q is intimately connected to the conservation of the Runge–Lenz vector **A** for the unperturbed hydrogen atom: the eigenvalue of the operator **A** is q/n—see, e.g., the textbook [8].

The second step of the analytical method of separating rapid and slow subsystems is to proceed to the slow subsystem (the CM motion), for which $E(\mathbf{R})$ from equation (10.15) will play the role of an effective potential. The effective Hamiltonian

$H_{CM,eff}(\mathbf{R}, \mathbf{P})$ for the CM motion becomes (the first, $\mathbf{R}$-independent term in $E(\mathbf{R})$ has been omitted because it does not affect the CM motion)

$$H_{CM,eff}(\mathbf{R}, \mathbf{P}) = P^2/[2(m_1 + m_2)] + (e_1 + e_2)\varphi(\mathbf{R}) \\ - (3n|q|\hbar^2/2)[1/(m_1 e_2) + 1/(m_2\, |e_1|)]F(\mathbf{R})\cos[\theta(\mathbf{R})]. \tag{10.16}$$

Thus, the application of this analytical method allowed the pseudoseparation of the CM motion and the relative motion for any two oppositely charged particles of significantly different masses in a nonuniform electric field.

It should be emphasized that in papers [5, 6], the CM coordinate $\mathbf{R}$ is considered as the *dynamical variable* (which generally depends on time) and that the Hamiltonian $H_{CM,eff}(\mathbf{R}, \mathbf{P})$ from equation (10.16) can be used to solve for the CM motion. This is the primary distinction of our work from papers where the CM coordinate of a hydrogenic atom/ion in a nonuniform electric field was considered to be fixed[1]. In section 3, we actually solve for the CM motion in the situation where the nonuniform electric field is due to the plasma ion nearest to the hydrogen atom, and apply the solution to the dynamical Stark broadening of hydrogen lines in plasmas. This would be impossible if the CM coordinate $\mathbf{R}$ was not treated as the *dynamical variable.*

We also note that higher order terms (quadrupole, octupole, etc) in the expansion of the potential $\varphi(\mathbf{R})$ in equation (10.4) can be easily taken into account, if necessary, and this analytical method for the pseudoseparation of the CM motion and the relative motion, with $\mathbf{R}$ considered as the *dynamical variable*, would still work.

In the particular case of hydrogen atoms one has

$$e_1 = e,\ e_2 = -e, \quad \mu = m_e m_p/(m_e + m_p), \tag{10.17}$$

where $e > 0$ is the electron charge, m_e and m_p are the electron and proton masses, respectively. Then equation (10.16) simplifies to

$$H_{CM,eff}(\mathbf{R}, \mathbf{P}) = P^2/(2m) - [3n|q|\hbar^2/(2\mu e)]F(\mathbf{R})\cos[\theta(\mathbf{R})], \\ m = (m_e + m_p). \tag{10.18}$$

Now we consider the situation where the nonuniform electric field is due to the nearest (to the hydrogen atom) ion of the positive charge Ze and mass m_i in a plasma located at the distance $\mathbf{R}$ from the hydrogen atom. Then the Hamiltonian from equation (10.18) can be rewritten as

[1] An example of such papers is Bekenstein and Krieger article [11]. In that article, while considering hydrogen atoms in a nonuniform electric field, the authors did not treat the CM coordinate as the dynamical variable, but rather considered it fixed, and did not provide any way to solve for the CM motion. We note in passing that among such papers on hydrogen atoms in a nonuniform electric field, Bekenstein and Krieger's article of 1970 [11] was in no way a pioneering work. Bekenstein and Krieger seemed unaware of Sholin's paper of 1969 [12], where he showed that the primary source of the asymmetry of hydrogen spectral line shapes is the nonuniformity of the plasma ion nearest to the hydrogen atom; moreover, Sholin took into account quadrupole and octupole interactions, while Bekenstein and Krieger allowed only for the quadrupole interaction.

$$H_{\mathrm{CM,eff}}(\mathbf{R}, \mathbf{P}) = P^2/(2m) - (D/R^2)\cos\theta, \\ D = [3n|q|\hbar^2/(2\mu)]Z, \quad \cos\theta = \mathbf{AR}/\mathrm{AR}. \tag{10.19}$$

This Hamiltonian represents a particle of mass m in the dipole potential. Since this particle is relatively heavy ($m \gg m_e$), its motion can be described classically and the corresponding classical solution is well-known—see, e.g., paper [13]. For this physical system, the radial motion can be exactly separated from the angular motion resulting in the following radial equation:

$$m[R(\mathrm{d}R/\mathrm{d}t) + (\mathrm{d}R/\mathrm{d}t)^2] = E_{\mathrm{CM}}, \tag{10.20}$$

where E_{CM} is the total energy of the particle. This equation allows the following exact general solution:

$$R(t) = (2E_{\mathrm{CM}}t^2/m + 2R_0v_0t + R_0^2)^{1/2}, \quad R_0 = R(0), \quad v_0 = (\mathrm{d}R/\mathrm{d}t)_{t=0}. \tag{10.21}$$

It is well-known that in plasmas of relatively low electron densities N_e, the Stark broadening of the most intense hydrogen lines, i.e., the lines corresponding to the radiative transitions between the levels of the low principal quantum numbers (such as, e.g., Ly-alpha, Ly-beta, H-alpha, etc), is dominated by the ion dynamical broadening—see, e.g., publications [14–20]. Let us discuss the corresponding validity condition in more detail.

The ion dynamical Stark broadening of hydrogen spectral lines is effective when the number ν_{Wi} of perturbing ions in the sphere of the ion Weisskopf radius is smaller than unity—see, e.g., review [9]. (In the opposite case of $\nu_{\mathrm{Wi}} \gg 1$, the perturbing ions can be treated in the quasistatic approximation.) By using the ion Weisskopf radius $R_{\mathrm{WA}}(\mathrm{C})$ defined in equation (10.50), one arrives at the following validity condition:

$$\nu_{\mathrm{Wi}}(C) = [3^{1/2}\pi/(2C^{3/2})](m_r/T)^{3/2}[(n^2 - n'^2)\hbar/\mu]^3 Z^2 N_e < 1. \tag{10.22}$$

For $C = 3/2$ (which is the choice of the strong collision constant in the conventional theory by Griem [21]) the numerical coefficient in the first brackets in the right side of equation (10.22) becomes $2^{1/2}\pi/3$. Thus, the ion dynamical Stark broadening can become effective for the most intense hydrogen spectral lines (i.e., for low values of n and n') in plasmas of relatively low electron densities.

Under the condition (10.22), for the overwhelming majority of perturbing ions, the frequency of the variation of the ion field v_i/R_N, where R_N is the mean interionic distance, exceeds the instantaneous Stark splitting in the ion field. Therefore the above requirement is called the modulation-type condition.

We note that there is another condition in Griem's book [21], his equation (82): $v_i/R_N > \gamma_e$, where γ_e is the electron impact width. This kind of requirement is called the damping-type condition. While both the modulation-type condition and the damping-type condition are necessary, the modulation-type condition (10.22) is more restrictive: it requires the electron (and ion) density to be by the factor $\sim (m_p/m_e)^{3/4} \sim 300$ smaller than the damping-type condition. Thus, the modulation-type condition overrides the damping-type condition.

In the so-called 'conventional theory' of the dynamical Stark broadening (also known as the 'standard theory') [21–23], the relative motion within the pair 'radiator—perturber' was assumed to occur along a straight line—as for a *free motion* (in our case the radiator is a hydrogen atom and the perturber is the perturbing ion). However, from the preceding discussion it follows that in the more advanced approach, the relative motion within the pair 'radiator–perturber' should be treated as the motion in the dipole potential—$(D/R^2)\cos\theta$, as seen from equation (10.19). The relevant setup of the problem is to choose the instant $t = 0$ as the instant of the smallest distance (the closest approach) within the pair 'radiator–perturber'. Then $v_0 = (dR/dt)_{t=0} = 0$, so that equation (10.21) simplifies to

$$R(t) = \left(2E_{CM}t^2/m + R_0^2\right)^{1/2}. \tag{10.23}$$

The energy E_{CM} can be represented in the form

$$E_{CM} = P_0^2/(2m) - \left(D/R_0^2\right)\cos\theta_0, \quad P_0 = P(0), \quad \theta_0 = \theta(0). \tag{10.24}$$

By considering the motion within the pair 'radiator–perturber' in the reference frame where the perturbing ion is at rest, so that $P_0 = mV_0$, where V_0 is the relative velocity within the pair 'radiator–perturber' at $t = 0$, the energy E_{CM} can be rewritten as

$$E_{CM} = mV_0^2/2 - \left(D/R_0^2\right)\cos\theta_0. \tag{10.25}$$

Then equation (10.22) becomes

$$R(t) = \left\{\left[V_0^2 - 2D\cos\theta_0/\left(mR_0^2\right)\right]t^2 + R_0^2\right\}^{1/2}. \tag{10.26}$$

By introducing the effective velocity

$$V_{eff}(R_0, \theta_0) = \left[V_0^2 - 2D\cos\theta_0/\left(mR_0^2\right)\right]^{1/2}, \tag{10.27}$$

we can make equation (10.26) formally equivalent to the usual case of the rectilinear trajectories:

$$R(t) = \left\{\left[V_{eff}(R_0, \theta_0)\right]^2 t^2 + R_0^2\right\}^{1/2}. \tag{10.28}$$

Now we consider a radiative transition between hydrogen energy levels a and b. In the general case, the ion dynamical broadening operator Φ_{ab} is defined as follows (by analogy with the electron dynamical broadening operator defined, e.g., in paper [24]):

$$\Phi_{ab}(t) = -\int dV_0 f(V_0) N_i V_0 \langle\sigma(V_0, \theta_0, t)\rangle_{\theta_0}. \tag{10.29}$$

Here $\langle\ldots\rangle_{\theta_0}$ denotes the averaging over the angle θ_0, and the operator $\sigma(V_0, \theta_0, t)$ has the form:

$$\sigma(V_0, \theta_0, t) = \int dR_0 2\pi R_0 [1 - U^{(R_0, V_0, \theta_0)}{}_a(t, 0) U^{(R_0, V_0, \theta_0)}{}_b{}^*(t, 0)]_{\text{ang.av}}. \tag{10.30}$$

Here N_i is the ion density, $f(V_0)$ is the distribution of the velocities (usually assumed to be Maxwellian), ρ is the impact parameter of the perturbing ion, U_a and U_b are the corresponding time-evolution operators, the symbols * and $[\ldots]_{\text{ang.av}}$ stand for the complex conjugation and the angular average, respectively. If time t were to be considered as a parameter, then the diagonal elements of the operator $\sigma(V_0, t)$ would have the physical meaning of cross-sections of so-called optical collisions, i.e., the cross-sections of collisions leading to virtual transitions inside level a between its sublevels and to virtual transitions inside level b between its sublevels, resulting in the broadening of Stark components of the hydrogen spectral line.

By using the trajectories from equation (10.26) and averaging over the polar angle θ_0, one can obtain the evolution operators and then the ion dynamical broadening operator with the allowance for the effect of the CM motion. However, in this general case, the results cannot be obtained analytically.

Therefore, for obtaining the final results analytically (which should help getting the message across in the simple form), we now employ the so-called impact approximation and substitute the evolution operators by the corresponding scattering matrices (see, e.g., papers [22, 23] or books [20, 21]):

$$\Phi_{ab} = -\int \mathrm{d}V_0 f(V_0) N_i V_0 \langle \sigma(V_0, \theta_0) \rangle_{\theta_0}, \tag{10.31}$$

$$\sigma(V_0, \theta_0) = \int \mathrm{d}R_0 2\pi R_0 [1 - S_a(R_0, V_0, \theta_0) S_b * (R_0, V_0, \theta_0)]_{\text{ang.av}}. \tag{10.32}$$

In the case where non-diagonal matrix elements of the Φ_{ab} are relatively small, the lineshape is a sum of Lorentzians, whose width $\gamma_{\alpha\beta}$ and shift $\Delta_{\alpha\beta}$ are equal (apart from the sign) to the real and imaginary parts of diagonal matrix elements $\langle\alpha|\langle\beta|\Phi_{ab}|\beta\rangle|\alpha\rangle$, respectively:

$$\gamma_{\alpha\beta} = -\mathrm{Re}[{}_{\alpha\beta}(F_{\text{ab}})_{\beta\alpha}],\ \Delta_{\alpha\beta} = -\mathrm{Im}[{}_{\alpha\beta}(F_{\text{ab}})_{\beta\alpha}]. \tag{10.33}$$

Here α and β correspond to upper and lower sublevels of the levels a and b, respectively. Here and below, for any operator G, for brevity we denote its matrix elements $\langle\alpha|\langle\beta|G|\beta\rangle|\alpha\rangle$ as ${}_{\alpha\beta}G_{\beta\alpha}$.

As we calculate the scattering matrices by the standard time-dependent perturbation theory, we obtain the following expression for the operator σ

$$\sigma(R_0, V_0, \theta_0) = \int \mathrm{d}R_0 2\pi R \left[K^2 Q(R_0, V_0, \theta_0)/R_0^2\right]. \tag{10.34}$$

Here

$$\begin{aligned} Q(R_0, V_0, \theta_0) &= 2\hbar^2/[3\mu^2 V_{\text{eff}}(R_0, \theta_0)^2] \\ &= Q_0/\left[1 - 2D\cos\theta_0/\left(mV_0^2 R_0^2\right)\right], \\ Q_0 &= 2Z^2\hbar^2/(3\mu^2 V_0^2), \end{aligned} \tag{10.35}$$

and

$$K^2 = K_a^{\ 2} + K_{\text{interf}} + K_b^{\ 2},\ K_a^{\ 2} = \mathbf{r}_a^{\ 2}/a_{\text{B}}^{\ 2},$$
$$K_{\text{interf}} = -\,2\mathbf{r}_a\mathbf{r}_b{}^*/a_{\text{B}}^{\ 2},\ K_b^{\ 2} = \mathbf{r}_b{}^{*2}/a_{\text{B}}^{\ 2},\ a_{\text{B}} = \hbar^2/(\mu e^2), \tag{10.36}$$

where a_{B} is the Bohr radius, K_{interf} represents the so-called interference term. In the conventional theory [21–23], in equation (10.35) instead of $V_{\text{eff}}(R_0,\ \theta_0)$, it would be $V_0^{\ 2}$.

The next step is the averaging of $1/V_{\text{eff}}(R_0,\ \theta_0)^2$ in equation (10.35) over the angle θ_0:

$$(1/2)\int_{-1}^{1} \text{d}(\cos\theta_0)\Big/\Big[V_0^2 - 2D\cos\theta_0\Big/\big(mR_0^2\big)\Big]$$
$$= \Big[R_0^2\Big/\big(2R_D^2V_0^2\big)\Big]\ln\Big[\big(R_0^2 + R_D^2\big)\Big/\big(R_0^2 - R_D^2\big)\Big], \tag{10.37}$$

where

$$R_D = \Big[2D\Big/\big(mV_0^2\big)\Big]^{1/2}, \tag{10.38}$$

so that the quantity $Q(R_0,\ \theta_0)$ after the averaging over θ_0 becomes

$$Q(R_0) = \Big[Q_0R_0^2\Big/\big(2R_D^2\big)\Big]\ln\Big[\big(R_0^2 + R_D^2\big)\Big/\big(R_0^2 - R_D^2\big)\Big] \tag{10.39}$$

with Q_0 defined in equation (10.35).

The way the quantity D (entering equation (10.38) was defined in equation (10.19) as $D = [3n|q|\hbar^2/(2\mu)]\ Z$ is valid only for the Lyman lines. For all other hydrogen lines one should use the arithmetic average of the values of D for the upper and lower Stark sublevels—as suggested in the similar case in paper [25] and used in paper [26]. Therefore, in the present paper for all other hydrogen lines we use the following value of D

$$D = 3(n|q| + n'|q'|)Ze^2a_B/4, \tag{10.40}$$

where the quantum numbers with the prime symbol and without it relate to the lower and upper levels, respectively.

The next step is the averaging over R_0. The integral over R_0 in equation (10.34) has a weak, logarithmic divergence at both small and large impact parameters—just like in the conventional theory [21–23]. Therefore, as in the conventional theory, we subdivide collisions into 'weak' ($R_0 > R_{\text{min}}$) and 'strong' ($R_0 < R_{\text{min}}$), and introduce also the upper cutoff R_{max} (just as in the conventional theory) discussed later. Then the diagonal elements of the cross-section of optical collisions can be represented in the form

$${}_{\alpha\beta}(\sigma)_{\beta\alpha,D} = \int_{R_{\text{min}}}^{R_{\text{max}}} \text{d}R_0 2\pi R_0\Big[{}_{\alpha\beta}(K^2)_{\beta\alpha}Q(R_0)/R_0^2\Big] + \int_0^{R_{\text{min}}} \text{d}R_0 2\pi R_0 C, \tag{10.41}$$

where R_{min} is defined by the condition:

$$ {}_{\alpha\beta}(K^2)_{\beta\alpha} Q(R_{\min})/R_{\min}{}^2 = {}_{\alpha\beta}(K^2)_{\beta\alpha}[Q_0/(2R_D{}^2)] \times \ln\left[\left(R_{\min}{}^2 + R_D{}^2\right)/\left(R_{\min}{}^2 - R_D{}^2\right)\right] = C \quad (10.42) $$

(naturally, $R_{\min} > R_D$). Here and below the subscript 'D' in ${}_{\alpha\beta}(\sigma)_{\beta\alpha,D}$ signifies that this cross-section was obtained with the allowance for the CM motion. The constant C in equation (10.42) is called 'strong collision constant' in the conventional theory. It arises from the preservation of the unitarity of the S-matrices:

$$ |1 - S_a(R_0, V_0, \theta_0) S_b{}^*(R_0, V_0, \theta_0)| = C, \quad C \leqslant 2. \quad (10.43) $$

For example, according to Griem's book [21], page 43, his choice was $C = 3/2$. More details can be found in paper [27][2].

As for the upper cutoff $R_{\max}$, following the conventional theory we choose it as the Debye radius

$$ R_{\max} = R_{\text{Debye}} = [T/(4\pi e^2 N_e)]^{1/2}, \quad (10.44) $$

though more rigorously, it should have been $R_{\max} = \min(R_{\text{Debye}}, V_0/\Delta\omega)$, where $\Delta\omega$ is the detuning from the center of the spectral line; physically, the requirements $R_{\max} < V_0/\Delta\omega$ being the allowance for incomplete collisions).

By integrating analytically over R_0 in equation (10.41) and substituting into the result the expression for the strong collision constant C from equation (10.42) we obtain:

$$ {}_{\alpha\beta}(\sigma)_{\beta\alpha,D} = 2\pi{}_{\alpha\beta}(K^2)_{\beta\alpha} Q_0 \left\{ \ln\left[\left(R_{\max}{}^4 - R_D{}^4\right)^{1/4} / \left(R_{\min}{}^4 - R_D{}^4\right)^{1/4}\right] + \left[R_{\max}{}^2/(4R_D{}^2)\right] \ln\left[\left(R_{\max}{}^2 + R_D{}^2\right)/\left(R_{\max}{}^2 - R_D{}^2\right)\right] \right\}. \quad (10.45) $$

The boundary $R_{\min}$ between the weak and strong collisions in equation (10.45) is the solution of equation (10.42) with respect to $R_{\min}$:

$$ R_{\min} = R_D \left\{ \left[\exp\left(2CR_D{}^2/{}_{\alpha\beta}(K^2)_{\beta\alpha} Q_0\right) + 1\right] / \left[\exp(2CR_D{}^2/{}_{\alpha\beta}(K^2)_{\beta\alpha} Q_0) - 1\right] \right\}^{1/2}. \quad (10.46) $$

---

[2] On page 43 of his book [21], Griem explicitly chose 3/2 for the quantity $|1 - S_a(R_0, V_0, \theta_0)\, S_b{}^*(R_0, V_0, \theta_0)|$ that we denoted as C. To avoid any confusion we note that what Griem called 'strong collision term' was $C/2$. The extra factor 1/2 arises from the following integral for the strong collision term:

$$ (1/\rho_{\min}{}^2) \int_0^{r_{\min}} d\rho\, \rho C = C/2. $$

The next step is the averaging of several quantities from the above equations over Stark sublevels of the upper and lower levels, so that each of these quantities will have the unique value for the particular hydrogen spectral line. First, the square root of the averaged matrix element $(\langle\alpha|\langle\beta|K^2|\beta\rangle|\alpha\rangle)$ is asserted to be

$$[{}_{\alpha\beta}(K^2)_{\beta\alpha}]_{\mathrm{av}}{}^{1/2} = \left[{}_{\alpha}\left(K_a{}^2\right)_{\alpha}{}^{1/2} - {}_{\beta}\left(K_b{}^2\right)_{\beta}{}^{1/2} \right]_{\mathrm{av}} \tag{10.47}$$

following the conventional theory justification [21] that in this form it allows for the partial cancellation of terms in ${}_{\alpha\beta}(W^2)_{\beta\alpha}$ when the principal quantum number n' of the lower level is close to the principal quantum number n of the upper level. The diagonal elements of the operators $K_a{}^2$ and $K_b{}^2$ have the following form in the parabolic coordinates (see, e.g. [10, 28])

$$\begin{aligned} {}_{\alpha}\left(K_a{}^2\right)_{\alpha} &= (9/8)n^2(n^2 + q^2 - m^2 - 1), \\ {}_{\beta}\left(K_b{}^2\right)_{\beta} &= (9/8)n'^2(n'^2 + q'^2 - m'^2 - 1)]. \end{aligned} \tag{10.48}$$

The averaging over Stark sublevels (since $(q^2)_{\mathrm{av}} = (m^2)_{\mathrm{av}}$) results in the following leading term in the quantity $[{}_{\alpha\beta}(K^2)_{\beta\alpha}]_{\mathrm{av}}{}^{1/2}$:

$$[{}_{\alpha\beta}(K^2)_{\beta\alpha}]_{\mathrm{av}} = (9/8)(n^2 - n'^2). \tag{10.49}$$

We mention that the same result (10.49) can be obtained after the corresponding averaging in the spherical quantization.

We denote

$$R_{\mathrm{WA}}(C) = \{[{}_{\alpha\beta}(K^2)_{\beta\alpha}]_{\mathrm{av}} Q_0 C\}^{1/2} = (3C)^{1/2}(n^2 - n'^2)\hbar Z/(2\mu V_0). \tag{10.50}$$

This quantity has the meaning of the so-called Weisskopf radius: it is defined here more accurately than in the conventional theory by Griem [21] (which is why here and below the superscript 'A' stands for 'accurate')—see appendix B of paper [6].

The next quantity to be averaged over Stark sublevels of the upper and lower levels, so that it will have the unique value for the particular hydrogen spectral line, is the quantity D from equation (10.40). The result reads:

$$\langle D \rangle_{\mathrm{av}} = (n^2 + n'^2)Ze^2 a_B/4. \tag{10.51}$$

After substituting this into the definition of R_D in equation (10.38), we obtain:

$$\langle R_D \rangle_{\mathrm{av}} = [(n^2 + n'^2)Z/2]^{1/2}\hbar/(\mu V_0). \tag{10.52}$$

Thus, from equations (10.46), (10.50), and (10.52), we get the unique value $\langle R_{\min}\rangle_{\mathrm{av}}$ for the entire hydrogen spectral line:

$$\begin{aligned} &\langle R_{\min}\rangle_{\mathrm{av}} \\ &= \langle R_D\rangle_{\mathrm{av}} \left\{\left[\exp(2\langle R_D\rangle_{\mathrm{av}}{}^2/R_{\mathrm{WA}}(C)^2) + 1\right]\Big/\left[\exp(2\langle R_D\rangle_{\mathrm{av}}{}^2/R_{\mathrm{WA}}(C)^2) - 1\right]\right\}^{1/2}. \end{aligned} \tag{10.53}$$

As the last step we substitute $R_{\min}$ by $\langle R_{\min}\rangle_{\rm av}$ and R_D by $\langle R_D\rangle_{\rm av}$ in equation (10.45), and also introduce dimensionless parameters

$$w = \langle R_D\rangle_{\rm av}/R_{\max}, \quad b = \langle R_D\rangle_{\rm av}/R_{\rm WA}(C) \\ = [2C/(3Z)]^{1/2}(n^2 + n'^2)^{1/2}/(n^2 - n'^2). \tag{10.54}$$

By doing so, we finally obtain:

$$_{\alpha\beta}(\sigma)_{\beta\alpha,A,D} = 2\pi_{\alpha\beta}(K^2)_{\beta\alpha}Q_0\{\ \ln[(\exp(2b^2) - 1)^{1/2}(1/w^4 - 1)^{1/4}/2^{1/2}] \\ - b^2/2 + [1/(4w^2)]\ln[(1 + w^2)/(1 - w^2)]\}, \tag{10.55}$$

where Q_0 was defined in equation (10.35) and

$$_{\alpha\beta}(K^2)_{\beta\alpha} = (9/8)[n^2(n^2 + q^2 - m^2 - 1) \\ - 4nqn'q' + n'^2(n'^2 + q'^2 - m'^2 - 1)]. \tag{10.56}$$

From equation (10.54) it is seen that the ratio $Z^{1/2}b/C^{1/2}$ is just a combination of the principal quantum numbers n and n' specific for each hydrogen spectral line: it is independent of the temperature T and of the electron density N_e of the plasma. Since the strong collision constant $C \leqslant 2$, it follows from equation (10.54) that $b < 1$ always. It reaches maximum values for $n' = n - 1$, i.e., for the most intense hydrogen spectral line of each spectral series. Here are examples for the case where the charge of the perturbing ions is $Z = 1$. For the Balmer-alpha line (H_α) we get $b = 0.59C^{1/2}$. For the Paschen-alpha, Brackett-alpha, and higher alpha lines, the ratio $b/C^{1/2}$ rapidly approaches $1/3^{1/2} = 0.58$. For the Lyman lines the expression for the ratio $b/C^{1/2}$ should be $2/(3^{1/2}n)$ instead of equation (10.54), so that for the Lyman-alpha line one gets $b/C^{1/2} = 1/3^{1/2}$ since $n = 2$.

The other dimensionless parameter $w = \langle R_D\rangle_{\rm av}/R_{\max}$, which enters equation (10.55), significantly depends on plasma parameters. In the most frequent case, where $R_{\max}$ is equal to the Debye radius R_D (given in equation (10.38)), the parameter w can be expressed as follows

$$w = [2e\hbar/(\mu T)][(n^2 + n'^2)Zm_rN_e]^{1/2} \\ = 8.99 \times 10^{-10}[(n^2 + n'^2)ZN_em_r/m_p]^{1/2}/T, \tag{10.57}$$

where

$$m_r = (m_e + m_p)m_i/(m_e + m_p + m_i). \tag{10.58}$$

In the utmost right part of equation (10.57), the temperature T is in eV and the electron density N_e is in cm^{-3}. While deriving equation (10.56), the quantity $1/V_0$ in the expression for $\langle R_D\rangle_{\rm av}$ (given by equation (10.52)) was substituted by its average over the Maxwell distribution $\langle 1/V_0\rangle = [2m_r/(\pi T)]^{1/2}$—just as in the conventional theory [21]. For the Lyman-lines, the expression for w should be modified to

$$w = [e\hbar n/(\mu T)](2m_rZN_e)^{1/2} = 1.27 \times 10^{-9}\, n[ZN_em_r/m_p]^{1/2}/T, \tag{10.59}$$

For presenting the effect of the CM motion in the universal form, it is convenient to introduce the ratio of the cross-section ${}_{\alpha\beta}(\sigma)_{\beta\alpha,A,D}$ to the corresponding cross-section ${}_{\alpha\beta}(\sigma)_{\beta\alpha,G}$ from the conventional theory by Griem [21]. Since the parameter w in equation (10.57) was obtained by averaging over the Maxwell distribution of the velocities, then the ratio of the cross-sections is essentially the same as the ratio of widths $\gamma_{\alpha\beta,A,D}/\gamma_{\alpha\beta,G}$:

$$\begin{aligned}\text{ratio} &= {}_{\alpha\beta}(\sigma)_{\beta\alpha,A,D}/{}_{\alpha\beta}(\sigma)_{\beta\alpha,G} = \gamma_{\alpha\beta,A,D}/\gamma_{\alpha\beta},G \\ &= \{\ln[(\exp(2b^2)-1)^{1/2}(1/w^4-1)^{1/4}/2^{1/2}] - b^2/2 + [1/(4w^2)] \\ &\quad \times \ln[(1+w^2)/(1-w^2)]\}/\{\ln[b/(wC^{1/2})] + 0.356\}.\end{aligned} \tag{10.60}$$

The matrix element ${}_{\alpha\beta}(W^2)_{\beta\alpha}$ cancels out from this ratio, so that it becomes indeed a universal function of just two dimensionless parameters w and b applicable for any set of the five parameters N_e, T, n, n', and C.

Below we provide numerical examples for some laboratory and astrophysical plasmas where the allowance for the CM motion significantly affects the ion dynamical Stark width. The first example is edge plasmas of magnetic fusion machines (such as, e.g., tokamaks), characterized by the electron density $N_e = (10^{14}\text{–}10^{15})\ \text{cm}^{-3}$ and the temperature of one or few eV (see, e.g., review [29]). For these plasma parameters, the Stark broadening of the most intense hydrogen spectral lines (Ly-alpha, Ly-beta, H-alpha, etc) can be dominated by the ion dynamical broadening (see, e.g., [14–20]).

The second example is plasmas in the atmospheres of flare stars. They are characterized by practically the same range of plasma parameters as the edge plasmas of magnetic fusion machines—see, e.g., book [30] and paper [31].

For both the edge of magnetic fusion machines and the atmospheres of flare stars, for the H_α line emitted from a hydrogen plasma at $N_e = 5 \times 10^{14}\ \text{cm}^{-3}$ and $T = 1$ eV, the ratio from equation (10.60) yields 1.19 for $C = 2$ and 1.13 for $C = 3/2$. Figure 10.1 presents this ratio (for the H_α line emitted from a hydrogen plasma) versus the electron density N_e at $T = 1$ eV for $C = 2$ (solid line) and for $C = 3/2$ (dashed line). It is seen that the allowance for the CM motion increases the ion dynamical Stark width of the H_α line in these kinds of plasmas by up to (15–20)%.

Our third example relates to plasmas of radiofrequency discharges, such as, e.g., those studied in papers [32–34]. The plasma parameters, e.g., in the experiments [32, 33], are $N_e = 1.2 \times 10^{13}\ \text{cm}^{-3}$ and $T = (1850\text{–}2000)$ K, i.e., $T = (0.16\text{–}0.17)$ eV. For the H_α line emitted from such a hydrogen plasma, the ratio from equation (10.60) yields 1.18 for $C = 2$ and 1.13 for $C = 3/2$. Figure 10.2 presents this ratio (for the H_α line emitted from a hydrogen plasma) versus the electron density N_e at $T = 0.17$ eV for $C = 2$ (solid line) and for $C = 3/2$ (dashed line). It is seen that the allowance for the CM motion increases the ion dynamical Stark width of the H_α line in these kinds of plasmas by up to (15%–20%).

In summary, in paper [5, 6] there was studied the general problem whether the CM motion and the relative motion can be separated for hydrogenic atoms/ions in a nonuniform electric field. It was demonstrated that, strictly speaking, they cannot be separated. Then the author of papers [5, 6] used the approximate analytical method

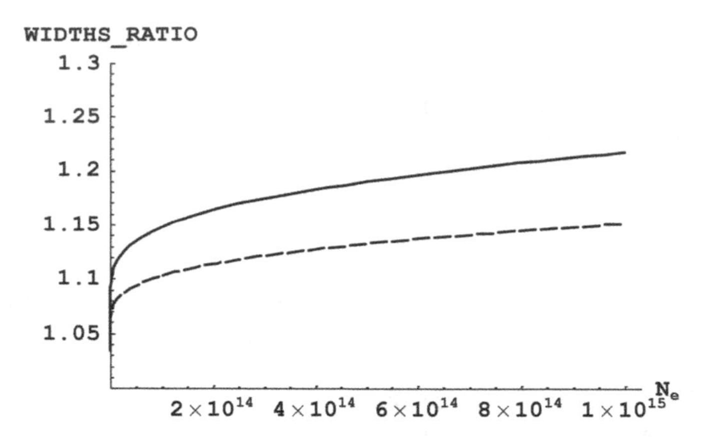

Figure 10.1. The ratio of the ion dynamical Stark width with the allowance for the CM motion to the ion dynamical Stark width from the conventional theory [21] versus the electron density N_e (cm^{-3}) for the H_α line emitted from a hydrogen plasma at $T = 1$ eV for $C = 2$ (solid line) and for $C = 3/2$ (dashed line). Plasma parameters correspond to edge plasmas of magnetic fusion machines and to atmospheres of flare stars. Reproduced with permission from [6]. Copyright 2018 E Oks.

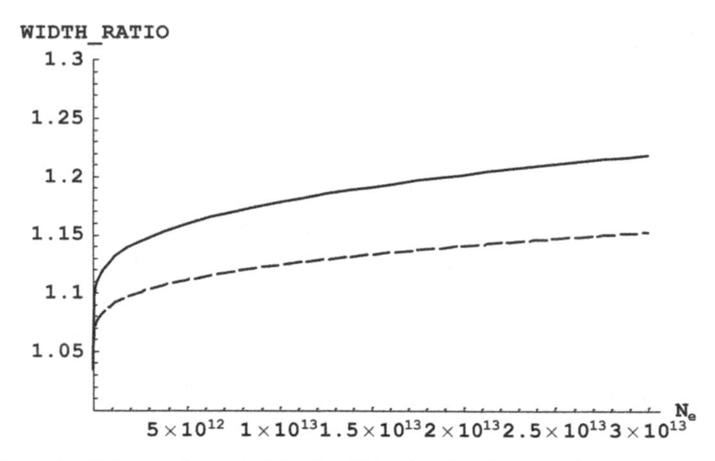

Figure 10.2. The ratio of the ion dynamical Stark width with the allowance for the CM motion to the ion dynamical Stark width from the conventional theory [21] versus the electron density N_e (cm^{-3}) for the H_α line emitted from a hydrogen plasma at $T = 0.17$ eV for $C = 2$ (solid line) and for $C = 3/2$ (dashed line). Plasma parameters correspond to radiofrequency discharges. Reproduced with permission from [6]. Copyright 2018 E Oks.

of the separation of rapid and slow subsystems to achieve the pseudoseparation of the CM and relative motions for hydrogenic atoms/ions in an arbitrary nonuniform electric field.

Next, in papers [5, 6] these results were further applied for the case of a hydrogen atom in the nonuniform electric field, where the field is due to the nearest (to the hydrogen atom) ion in a plasma. It was shown that the effect of the CM motion can be formally taken into account via the substitution of the initial relative velocity V_0

in the pair 'atom–ion' by an effective velocity V_{eff} that depends on the quantum numbers of the hydrogen atom, as well as on the initial separation R_0 in the pair 'atom–ion' and on the ion charge Z.

Finally, in papers [5, 6] the results were applied to the ion dynamical Stark broadening of hydrogen lines in plasmas. There were obtained analytical results for the cross-sections of the optical collisions that control the corresponding Stark width. There were presented specific examples of laboratory plasmas (such as magnetic fusion plasmas or plasmas of radiofrequency discharges) and astrophysical plasmas (such as in atmospheres of flare stars) where the allowance for these CM effects leads to a significant increase of the width of hydrogen spectral lines—by up to 15%–20%.

Thus, in addition to the fundamental importance, the results of papers [5, 6] have also practical importance for spectroscopic diagnostics of laboratory and astrophysical plasmas.

References

[1] Schmelcher P and Cederbaum L S 1995 *Phys. Rev. Lett.* **74** 662
[2] Schmelcher P and Cederbaum L S 1993 *Phys. Rev.* A **47** 2634
[3] Vincke M and Baye D 1988 *J. Phys. B: At. Mol. Opt. Phys.* **21** 2407
[4] Kotkin G L and Serbo V G 1971 *Collection of Problems in Classical Mechanics* (Oxford: Pergamon) problem 2.22
[5] Oks E 2017 *Inter. Rev. Atom. Mol. Phys.* **8** 41
[6] Oks E 2018 *J. Phys. Comm.* **2** 045005
[7] Galitski V, Karnakov B, Kogan V and Galitski V Jr 2013 *Exploring Quantum Mechanics* (Oxford: Oxford University Press), p 363
[8] Landau L D and Lifshitz E M 1965 *Quantum Mechanics* (Oxford: Pergamon)
[9] Lisitsa V S 1977 *Sov. Phys. Uspekhi.* **122** 603
[10] Lisitsa V S 1994 *Atoms in Plasmas* (Berlin: Springer), p 13
[11] Bekenstein J D and Krieger J B 1970 *J. Math. Phys.* **11** 2721
[12] Sholin G V 1969 *Opt. Spectrosc.* **26** 275
[13] Fox K 1968 *J. Phys. A (Proc. Phys. Soc.), ser.* **2** 124
[14] Abramov V A and Lisitsa V S 1977 *Sov. J. Plasma Phys.* **3** 451
[15] Seidel J 1979 *Z. Naturforsch.* **34a** 1385
[16] Stehle C and Feautrier N 1984 *J. Phys.* B **17** 1477
[17] Derevianko A and Oks E 1994 *Phys. Rev. Lett.* **73** 2059
[18] Derevianko A and Oks E 1995 *J. Quant. Spectr. Rad. Transf.* **54** 137
[19] Derevianko A and Oks E 1997 *Rev. Sci. Instrum.* **68** 998
[20] Oks E 2006 *Stark Broadening of Hydrogen and Hydrogenlike Spectral Lines in Plasmas: The Physical Insight* (Oxford: Alpha Science International)
[21] Griem H R 1974 *Spectral Line Broadening by Plasmas* (New York: Academic)
[22] Baranger M 1958 *Phys. Rev.* **111** 494
[23] Kepple P and Griem H R 1968 *Phys. Rev.* **173** 317
[24] Kolb A C and Griem H R 1958 *Phys. Rev.* **111** 514
[25] Nienhuis G 1973 *Physica* **66** 245
[26] Szudy J and Baylis W E 1976 *Canad. J. Phys.* **54** 2287

[27] Oks E 2015 *J. Quant. Spectr. Rad. Transf.* **152** 74
[28] Sholin G V, Demura A V and Lisitsa V S 1973 *Sov. Phys. JETP* **37** 1057
[29] Pospieszczyk A 2005 *Phys. Scripta* **2005** T119
[30] Gershberg R E 2005 *Solar-Type Activity in Main-Sequence Stars* (Berlin: Springer)
[31] Oks E and Gershberg R E 2016 *Astrophys. J.* **819** 16
[32] Bengston R D, Tannich J D and Kepple P 1970 *Phys. Rev.* A **1** 532
[33] Bengtson R D and Chester G R 1972 *Astrophys. J.* **178** 565
[34] Himmel G 1976 *J. Quant. Spectrosc. Rad. Transf.* **16** 529

Chapter 11

Advanced treatment of the Stark broadening of hydrogen spectral lines by plasma electrons

The Stark broadening (SB) of hydrogen lines (H-lines, including deuterium and tritium lines) is important for many applications in plasma physics (including controlled fusion and plasma processing), in laser-induced breakdown spectroscopy, and in astrophysics. The most 'user-friendly' are semiclassical theories of the SB of H-lines—because their results can be expressed analytically in a relatively simple form for any H-line.

Within the semiclassical theories, the simplest is the so-called conventional theory (CT)—also known as the 'standard theory'. In the CT, it is assumed that from the viewpoint of the radiator, the ion microfield (i.e., the electric field due to ionic perturbers) is quasistatic, while the electron microfield is treated dynamically in the so-called impact approximation [1, 2]. (The CT is frequently referred to as Griem's theory—as presented in the Kepple–Griem paper [2] and in Griem's book [3].) A more accurate version of the collisional theory of the SB is called the unified theory. Physically, the primary distinction of the unified theory from the CT is the allowance for incomplete collisions [4, 5]. A significant analytical advance within the CT was presented in papers [6, 7], where it was found that the Stark profile of hydrogenic spectral lines caused by the electrons at the absence of the ion microfield reduces to a single Lorentzian. Analytical advances going beyond the CT have been exact (non-perturbative—in distinction to the CT) analytical solutions of the problem—but only if all 'perturbing' charges were of the same sort [8, 9].

The next advance in analytical theories of the SB beyond the CT was called generalized theory (GT). It demonstrated for the first time that the coupling of the electron and ion microfields can be strong [10, 11]—see also book [12], chapter 4. (In paper [13] there was previously found only a weak, logarithmic coupling between the electron and ion microfields.) This *indirect coupling* (facilitated by the radiator) increases with the growth of the electron density N_e and/or the principal quantum number n, as well as with the decrease of the temperature T [10–12]. The GT accomplished this by going beyond the fully-perturbative description of the electron microfield used in the CT.

Griem's version of the CT presented in Kepple–Griem paper [2] is still used by a number of groups performing laboratory experiments or astrophysical observations (especially, by the latter groups) for comparison with their experimental or observational results. Therefore, in papers [14, 15] the author came back to the CT for H-lines (still using the impact approximation for electrons and the quasistatic approximation for ions, just as in Griem's CT) and develop a refined CT by taking into account the following. The characteristic frequency of the variation of the perturbing electron electric field at the location of the radiating atom is much smaller than the atomic transition frequency (speaking quantally) or the Kepler frequency of the atomic electron (speaking quasi-classically). So, the perturbing electron represents a slow subsystem while the radiation atom represents the rapid subsystem. After averaging over the rapid subsystem, the perturbing electron moves in the effective potential caused by the fact that hydrogen atoms possess permanent dipole moments (in the overwhelming majority of the atomic states). This fact is intimately related to the existence of an additional conserved vector quantity—the Runge–Lenz vector **A** (discussed, e.g., in textbook [16]).

The average value of the dipole moment is antiparallel to the vector **A** (see, e.g., [17–19]). Therefore perturbing electrons move in a dipole potential

$$V = e^2\langle \mathbf{R}\rangle \bullet \mathbf{r} / \mathbf{r}^3, \tag{11.1}$$

where **r** is the radius-vector of the perturbing electrons and $\langle \mathbf{R}\rangle$ is the mean value of the radius-vector of the atomic electron:

$$\langle \mathbf{R}\rangle = -3e^2\mathbf{A}/(4\,|E_{\text{at}}|), \tag{11.2}$$

where E_{at} is the energy of the atomic electron. Hence the perturbing electrons actually do not move as free particles—in distinction to Griem's CT.

Another refinement of the CT in papers [14, 15] had to do with the fact that Griem's definition of the so-called Weisskopf radius (defined in the next section) was not quite accurate. Also, in his book [3] Griem suggested changing so-called strong collision constant (defined in the next section) without changing the Weisskopf radius, while in reality the choices of the Weisskopf radius and of the strong collision constant are interrelated: changing the strong collision constant necessitates the corresponding change of the Weisskopf radius.

One of the primary analytical results from papers [14, 15] is the following ratio of Stark widths calculated with ($\gamma_{\alpha\beta,A,d}$) and without ($\gamma_{\alpha\beta,A}$) the allowance for the scattering of the perturbing electrons on the atomic electric dipole (both $\gamma_{\alpha\beta,A,d}$ and $\gamma_{\alpha\beta,A}$ being calculated using the more accurate definition of the Weisskopf radius, which is why they have the subscript A):

$$\begin{aligned}\text{ratio} = \gamma_{\alpha\beta,A,d}/\gamma_{\alpha\beta,A} = \{&\ln[(\exp(2b^2) - 1)^{1/2}(1/x^4 - 1)^{1/4}/2^{1/2}] - b^2/2 \\ &+ [1/(4x^2)]\ln[(1 + x^2)/(1 - x^2)]\}/\{\ln[b/x] + 1/2\}.\end{aligned} \tag{11.3}$$

Here, α and β correspond to upper and lower sublevels of the levels a and b involved in the radiative transition, respectively. In equation (11.3)

$$b = (2C/3)^{1/2}(n^2 + n'^2)^{1/2}/(n^2 - n'^2), \quad (11.4)$$

where n and n' are the principal quantum numbers of the upper and lower energy levels, involved in the radiative transition, respectively; C is the strong collision constant discussed below, and

$$x = (2e\hbar/T)[(n^2 + n'^2)N_e/m_e]^{1/2} = 2.097 \times 10^{-11}\,[(n^2 + n'^2)N_e]^{1/2}/T, \quad (11.5)$$

where in the last, 'practical' part of equation (11.5), the temperature T is in eV and the electron density N_e is in cm^{-3}.

The strong collision constant in the CT shows up in the condition of the unitarity of the scattering matrices S_a and S_b

$$|1 - S_a(\rho, v)\, S_b{}^*(\rho, v)| = C, \quad C \leqslant 2, \quad (11.6)$$

where ρ and v are the impact parameter and the velocity of the perturbing electron, respectively; the symbol * stands for the complex conjugation.

For comparison with Griem's CT, a modification has to be made to the corresponding ratio of widths. After taking into account that Griem's choice of the Weisskopf radius, the corresponding ratio becomes

$$\begin{aligned}\text{ratio } A \text{ to } G = \gamma_{\alpha\beta,A,d}/\gamma_{\alpha\beta,G} = \{&\ln[(\exp(2b^2) - 1)^{1/2}(1/x^4 - 1)^{1/4}/2^{1/2}] - b^2/2 \\ &+ [1/(4x^2)]\ln[(1 + x^2)/(1 - x^2)]\}/\{\ln[b/(xC^{1/2})] \\ &+ 0.356\}.\end{aligned} \quad (11.7)$$

Figure 11.1 presents the ratio of the Stark widths from equation (11.7) for the H_α line versus the dimensionless parameter x (given by equation (11.5)) for three different choices of the strong collision constant: $C = 2$ (solid curve), $C = 3/2$ (dotted curve), as suggested on page 70 of Griem's book [3], and $C = 1$ (dashed curve).

It is seen that in the more advanced version of the CT, the Stark width of the H_α line can exceed the corresponding width from Griem's CT by up to 150%. We note that the corresponding figure 2 from papers [14, 15], presenting the same ratio A to G for $C = 1$ and $C = 2$, is regrettably erroneous.

Figures 11.2 and 11.3, corresponding to figures 3 and 4 from papers [14, 15], present the comparison of the experimental widths of the H_α line from two different benchmark experiments with several theories. Namely, in figure 11.2 the experimental data was obtained by Kunze's group in a gas-liner pinch plasma [20], while in figure 11.3 the experimental data was obtained by Vitel's group in a flash tube plasma. In both figures the experimental widths are presented by separated dots. As for the theories, in both figures the solid curve corresponds to the refined CT from the present paper, the dotted curve—to Griem's CT, and the dashed curve—to the generalized theory (GT). It should be emphasized that the theoretical widths based on the refined CT, developed in papers [14, 15], have been calculated taking into account both diagonal and nondiagonal matrix elements of the electron impact

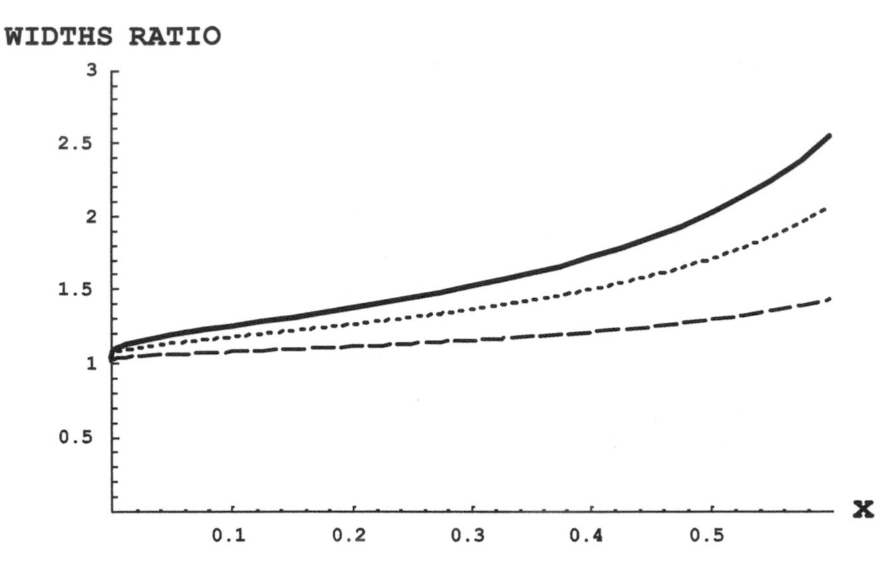

Figure 11.1. The ratio of the Stark widths from equation (11.7) for the H_α line versus the dimensionless parameter x (given by equation (11.5)) for three different choices of the strong collision constant: $C = 2$ (solid curve), $C = 3/2$ (dotted curve), as suggested on page 70 of Griem's book [3], and $C = 1$ (dashed curve).

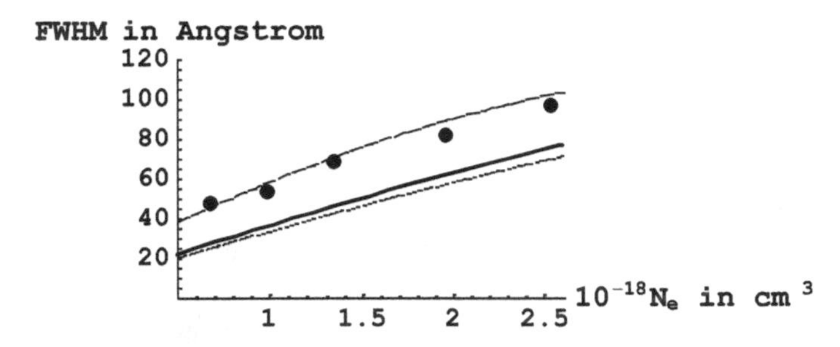

Figure 11.2. Comparison of the experimental widths of the H_α line (separated dots) obtained by Kunze's group in a gas-liner pinch plasma [20] at the temperatures (6–8) eV with the following theories: the refined CT from the present paper (solid line), Griem's CT (dotted line), the GT (dashed line). (Reprinted by permission from [15]. Copyright 2015 Elsevier.)

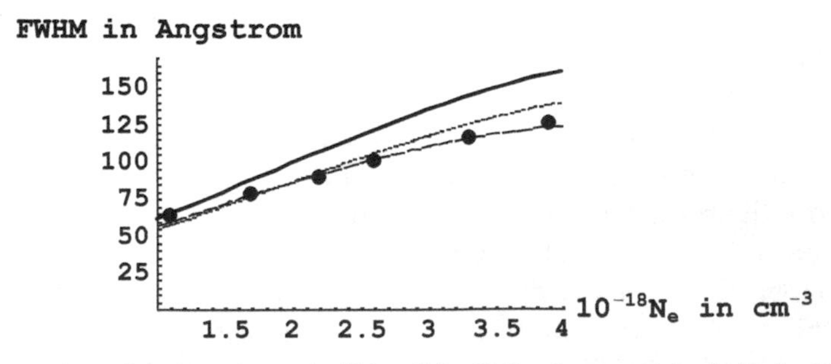

Figure 11.3. Comparison of the experimental widths of the H_α line (separated dots) obtained by Vitel's group in a flash tube plasma [21] at the temperatures (1–1.5) eV and the initial gas pressure 600 Torr with the following theories: the refined CT from the present paper (solid line), Griem's CT (dotted line), the GT (dashed line). (Reprinted by permission from [15]. Copyright 2015 Elsevier.)

broadening operator, as well as the quasistatic broadening by ions, i.e., in the same way as in Griem's CT. We also note that the theoretical widths based on the GT were not a new result, but rather were those that had been previously calculated and presented in book [12] in sections 9.1 and 9.2.

Thus, the above refinements of the CT increase the electron broadening. This is especially evident for warm dense plasmas emitting H-lines.

References

[1] Baranger M 1958 *Phys. Rev.* **111** 481

[2] Kepple P and Griem H R 1968 *Phys. Rev.* **173** 317

[3] Griem H R 1974 *Spectral Line Broadening by Plasmas* (New York: Academic)

[4] Smith E, Cooper J and Vidal C 1969 *Phys. Rev.* **185** 140

[5] Vidal C, Cooper J and Smith E 1970 *J. Quant. Spectrosc. Rad. Transf.* **10** 1011

[6] Stehle C and Feautrier N 1984 *J. Phys. B: Atom. Mol. Phys.* **17** 1477

[7] Stehle C 1995 *Astron. & Astrophys.* **305** 677

[8] Lisitsa V S and Sholin G V 1972 *Sov. Phys. JETP* **34** 484

[9] Derevianko A and Oks E 1996 *Physics of Strongly Coupled Plasmas* ed W D Kraeft and M Schlanges (Singapore: World Scientific), p 286

[10] Ispolatov Y and Oks E 1994 *J. Quant. Spectrosc. Rad. Transf.* **51** 129

[11] Oks E, Derevianko A and Ispolatov Y 1995 *J. Quant. Spectrosc. Rad. Transf.* **54** 307

[12] Oks E 2006 *Stark Broadening of Hydrogen and Hydrogenlike Spectral Lines in Plasmas: The Physical Insight* (Oxford: Alpha Science International)

[13] Sholin G V, Demura A V and Lisitsa V S 1973 *Sov. Phys. JETP* **37** 1057

[14] Oks E 2013 *Intern. Rev. Atom. Mol. Phys.* **4** 49

[15] Oks E 2015 *J. Quant. Spectrosc. Rad. Transf.* **152** 74

[16] Landau L D and Lifshitz E M 1965 *Quantum Mechanics* (Oxford: Pergamon)

[17] Kotkin G L and Serbo V G 1971 *Collection of Problems in Classical Mechanics* (Oxford: Pergamon) problem 3.32

[18] Lisitsa V S 1977 *Sov. Phys. Uspekhi* **122** 603

[19] Lisitsa V S 1994 *Atoms in Plasmas* (Berlin: Springer), p 13

[20] Büscher S, Wrubel T, Ferri S and Kunze H J 2002 *J. Phys. B: Atom. Mol. Opt. Phys.* **35** 2889

[21] Flih S A, Oks E and Vitel Y 2003 *J. Phys. B: Atom. Mol. Phys.* **36** 283

Chapter 12

Advanced treatment of the Stark broadening of hydrogen-like spectral lines by plasma electrons

The theory of the Stark broadening of hydrogen-like spectral lines by plasma electrons, developed by Griem and Shen [1] and later presented also in books [2, 3], is usually referred to as the conventional theory, hereafter CT, also known as the standard theory. (Further advances in the theory of the Stark broadening of hydrogen-like spectral lines by plasma electrons can be found, e.g., in books [4, 5] and references therein.) In the CT, the perturbing electrons are considered moving along hyperbolic trajectories in the Coulomb field of the effective charge $Z - 1$ (in atomic units), where Z is the nuclear charge of the radiating ion. In other words, in the CT there was made a simplifying assumption that the motion of the perturbing electron can be described in frames of a two-body problem, one particle being the perturbing electron and the other 'particle' being the charge $Z - 1$.

However, in reality one has to deal with a three-body problem: the perturbing electron, the nucleus, and the bound electron. Therefore, trajectories of the perturbing electrons should be more complicated.

In paper [6], the authors took this into account by using the standard analytical method of separating rapid and slow subsystems—see, e.g., book [7]. The characteristic frequency of the motion of the bound electron around the nucleus is much higher than the characteristic frequency of the motion of the perturbing electron around the radiating ion. Therefore, the former represents the rapid subsystem and the latter represents the slow subsystem. This approximate analytical method allows a sufficiently accurate treatment in situations where the perturbation theory fails—see, e.g., book [7].

By applying this method the authors of paper [6] obtained more accurate analytical results for the electron broadening operator than in the CT. They showed by examples of the electron broadening of the Lyman lines of He II that the allowance for this effect increases with the electron density N_e, becomes significant

doi:10.1088/2053-2563/ab3db0ch12

already at $N_e \sim 10^{17}$ cm^{-3} and very significant at higher densities. Here are the details.

In the CT the electron broadening operator is expressed in the form (see, e.g., paper [1])

$$\Phi_{ab} \equiv 2\pi v N_e \int d\rho \rho \{S_a S_b^* - 1\}. \tag{12.1}$$

where N_e, v, and ρ are the electron density, velocity, and impact parameter, respectively; $S_a(0)$ and $S_b(0)$ are the S matrices for the upper (a) and lower (b) states involved in the radiative transition, respectively; $\{\ldots\}$ stands for the averaging over angular variables of vectors $\mathbf{v}$ and $\boldsymbol{\rho}$. Further in the CT, the collisions are subdivided into weak and strong. The weak collisions are treated by the time-dependent perturbation theory. The impact parameter, at which the formally calculated expression $\{S_a S_b^* - 1\}$ for a weak collision starts violating the unitarity of the S-matrices, serves as the boundary between the weak and strong collisions and is called Weisskopf radius ρ_{We}.

So, in the CT the integral over the impact parameter diverges at small ρ. Therefore, in the CT this integral is broken down into two parts: from 0 to ρ_{We} (strong collisions) and from ρ_{We} to ρ_{max} for weak collisions. The upper cutoff ρ_{max} (typically chosen to be the Debye radius $\rho_{\mathrm{D}} = [T/(4\pi e^2 N_e)]^{1/2}$, where T is the electron temperature) is necessary because this integral diverges also at large ρ.

In the CT, after calculating the S matrices for weak collisions, the electron broadening operator becomes (*in atomic units*)

$$\Phi_{ab}{}^{\mathrm{weak}} \equiv C \int_{\rho_{\mathrm{we}}}^{\rho_{\mathrm{max}}} d\rho \rho \sin^2 \frac{\Theta(\rho)}{2} = \frac{C}{2} \int_{\Theta_{\mathrm{min}}}^{\Theta_{\mathrm{max}}} d\Theta \frac{d\rho^2}{d\Theta} \sin^2 \frac{\Theta}{2}, \tag{12.2}$$

where Θ is the scattering angle for the collision between the perturbing electron and the radiating ion (the dependence between Θ and ρ being discussed below) and the plasma electron and the operator C is

$$C = -\frac{4\pi}{3} N_e \left[\int_0^{\infty} dv v^3 f(v) \right] \frac{m^2}{(Z-1)^2} (\boldsymbol{r}_a - \boldsymbol{r}_b{}^*)^2. \tag{12.3}$$

Here $f(v)$ is the velocity distribution of the perturbing electrons, $\mathbf{r}$ is the radius-vector operator of the bound electron (which scales with Z as $1/Z$), and m is the reduced mass of the system 'perturbing electron—radiating ion'.

In the CT the scattering occurs in the effective Coulomb potential, so that the trajectory of the perturbing electron is hyperbolic and the relation between the impact parameter and the scattering angle is given by

$$\rho^{(0)} = \frac{Z-1}{mv^2} \cot \frac{\Theta}{2}. \tag{12.4}$$

In paper [6] for a more realistic description of the situation, the authors used the standard analytical method of separating rapid and slow subsystems [6], as noted above. It is applicable here because the characteristic frequency $v_{\mathrm{Te}}/\rho_{\mathrm{We}}$ of the variation the electric field of the perturbing electrons at the location of the radiating ion is much smaller than the frequency Ω_{ab} of the spectral line (the latter, e.g., in case of the radiative transition between the Rydberg states would be the Kepler frequency or its harmonics).

Indeed, the characteristic frequency of the motion of the perturbing electron around the radiating ion in the process of the Stark broadening of spectral lines is the so-called Weisskopf frequency

$$\omega_{\mathrm{We}} = \frac{v_T}{\rho_{\mathrm{We}}} \sim \frac{Zmv_T^2}{\left(n_a^2 - n_b^2\right)\hbar} \sim \frac{ZT}{\left(n_a^2 - n_b^2\right)\hbar}. \tag{12.5}$$

The characteristic frequency of the motion of the bound electron around the nucleus is the frequency of the spectral line

$$\Omega = \frac{Z^2 U_{\mathrm{H}}}{\hbar}\left(\frac{1}{n_b^2} - \frac{1}{n_a^2}\right), \tag{12.6}$$

where U_{H} is the ionization potential of hydrogen. The ratio of these two frequencies is

$$\frac{\omega_{\mathrm{We}}}{\Omega} \sim \left(\frac{T}{ZU_{\mathrm{H}}}\right)\left[\frac{n_a^2 n_b^2}{\left(n_a^2 - n_b^2\right)^2}\right]. \tag{12.7}$$

For the simplicity of estimating this ratio, let us consider $n_a \gg n_b$, so that

$$\frac{\omega_{\mathrm{We}}}{\Omega} \sim \left(\frac{T}{Zn_a^2 U_{\mathrm{H}}}\right) \ll 1 \tag{12.8}$$

as long as

$$T(\mathrm{eV}) \ll (13.6\ \mathrm{eV})Zn_a^2. \tag{12.9}$$

For

$$T(\mathrm{eV}) \ll (27.2\ \mathrm{eV})n_a^2 \tag{12.10}$$

and is satisfied for a broad range of temperatures, at which He II spectral lines are observed in plasmas.

The first step in this method is to 'freeze' the slow subsystem (perturbing electron) and to find the analytical solution for the energy of the rapid subsystem (the radiating ion) that would depend on the frozen coordinates of the slow subsystem (in our case it will be the dependence on the distance R of the perturbing electron from the radiating ion). To the first non-vanishing order of the R-dependence, the corresponding energy in the parabolic quantization is given by

$$E_{nq}(R) = -\frac{Z^2}{n^2} + \frac{3nq}{2ZR^2}, \tag{12.11}$$

where n and $q = n_1 - n_2$ are the principal and electric quantum numbers, respectively; n_1 and n_2 are the parabolic quantum numbers.

The next step in this method is to consider the motion of the slow subsystem (perturbing electron) in the 'effective potential' $V_{\text{eff}}(R)$ consisting of the actual potential plus $E_{\text{nq}}(R)$. Since the constant term in equation (12.11) does not affect the motion, the effective potential for the motion of the perturbing electron can be represented in the form

$$V_{\text{eff}}(R) = -\frac{\alpha}{R} + \frac{\beta}{R^2}, \ \alpha = Z - 1. \tag{12.12}$$

For the spectral lines of the Lyman series, since the lower (ground) state b of the radiating ion remains unperturbed (up to/including the order $\sim 1/R^2$), the coefficient β is

$$\beta = \frac{3n_a q_a}{2Z}. \tag{12.13}$$

For other hydrogenic spectral lines, for taking into account both the upper and lower states of the radiating ion, the coefficient β can be expressed as

$$\beta = \frac{3(n_a q_a - n_b q_b)}{2Z}. \tag{12.14}$$

The motion in the potential from equation (12.12) allows an exact analytical solution. In particular, the relation between the scattering angle and the impact parameter is no longer given by equation (12.4), but rather becomes (see, e.g., book [8])

$$\Theta = \pi - \frac{2}{\sqrt{1 + \dfrac{2m\beta}{M^2}}} \arctan \sqrt{\frac{4E}{\alpha^2}\left(\beta + \frac{M^2}{2m}\right)}. \tag{12.15}$$

Here, E and M are the energy and the angular momentum of the perturbing electron, respectively. We can rewrite the angular momentum in terms of the impact parameter ρ as

$$M = mv\rho \tag{12.16}$$

Then a slight rearrangement of equation (12.15) yields

$$\tan\left(\frac{\pi - \Theta}{2}\sqrt{1 + \frac{2\beta}{mv^2\rho^2}}\right) = \frac{v}{\alpha}\sqrt{m^2v^2\rho^2 + 2m\beta}. \tag{12.17}$$

After solving equation (12.17) for ρ and substituting the outcome in equation (12.2), a more accurate expression for the electron broadening operator can be obtained. However, equation (12.17) does not have an exact analytic solution for ρ so that this could be done only numerically.

In the paper [6] for getting the message across in the simplest form, the authors provided an approximate analytical solution of equation (12.17) by expanding it in powers of β. This yields (keeping up to the first power of β)

$$\tan\left(\frac{\pi-\Theta}{2}\right)+\left(\frac{\pi-\Theta}{2}\right)\left[1+\tan^2\left(\frac{\pi-\Theta}{2}\right)\right]\frac{\beta}{mv^2\rho^2}\approx\frac{mv^2\rho}{\alpha}+\frac{\beta}{\alpha\rho}. \tag{12.18}$$

The analytical solution for ρ was sought in the form $\rho \approx \rho^{(0)} + \rho^{(1)}$, where $\rho^{(0)}$ corresponds to $\beta = 0$ (and was given by equation (12.4)) and $\rho^{(1)} \ll \rho^{(0)}$. Substitution of $\rho \approx \rho^{(0)} + \rho^{(1)}$ into equation (12.18) yields the expression

$$\frac{(\pi-\Theta)\beta}{2mv^2\rho^{(0)2}\sin^2\dfrac{\Theta}{2}}-\frac{\beta}{\alpha\rho^{(0)}}\approx\frac{mv^2\rho^{(1)}}{\alpha}. \tag{12.19}$$

After solving equation (12.19) for $\rho^{(1)}$, the following expression for ρ was obtained in paper [6]:

$$\rho\approx\frac{\alpha}{mv^2}\cot\frac{\Theta}{2}+\frac{\beta}{\alpha}\left(\frac{\pi-\Theta}{2\cos^2\dfrac{\Theta}{2}}-\tan\frac{\Theta}{2}\right). \tag{12.20}$$

As a reminder, the goal is to perform the integration in equation (12.1) for obtaining a more accurate analytical result for the electron broadening operator. This can be more easily accomplished by performing the integration over Θ instead of ρ. For this purpose, first the authors of paper [6] squared equation (12.20)

$$\rho^2\approx\frac{\alpha^2}{m^2v^4}\cot^2\frac{\Theta}{2}+\frac{\beta}{mv^2}\left(\frac{\pi-\Theta}{\sin\dfrac{\Theta}{2}\cos\dfrac{\Theta}{2}}-1\right). \tag{12.21}$$

where only the first-order terms in β have been kept for consistency. To make formulas simpler, they denoted $\phi = \Theta/2$. After differentiating equation (12.21) with respect to ϕ, they obtained

$$\begin{aligned}\frac{d\rho^2}{d\phi}\approx&-\frac{\alpha^2}{m^2v^4}\frac{2\cot\phi}{\sin^2\phi}-\frac{2\beta}{mv^2}\\&\times\left[\left(\frac{1}{\sin\phi\cos\phi}\right)+\left(\frac{\pi}{2}-\phi\right)\left(\frac{1}{\sin^2\phi}-\frac{1}{\cos^2\phi}\right)\right]\end{aligned} \tag{12.22}$$

After substituting in the utmost right side of equation (12.2) first $\Theta = 2\phi$ and then $\frac{d\rho^2}{d\phi}$ from equation (12.22), the contribution of the weak collisions to the electron broadening operator becomes

$$\Phi_{ab}{}^{\text{weak}} = -C\left[\frac{\alpha^2}{m^2v^4}\int_{\phi_{\min}}^{\phi_{\max}} \cot\phi\, d\phi + \frac{\beta}{mv^2}\int_0^{\frac{\pi}{2}} \tan\phi\, d\phi \right.$$
$$\left. + \frac{\beta}{mv^2}\int_0^{\frac{\pi}{2}} \left(\frac{\pi}{2} - \phi\right)(1 - \tan^2\phi)d\phi\right]. \tag{12.23}$$

In equation (12.23), in the two correction terms proportional to β, the authors of paper [6] extended the integration over the full range of the variation of the angle ϕ. The corresponding minor inaccuracy would not contribute significantly to the electron broadening operator, since the terms involving β are considered to be a relatively small correction to the first term in equation (12.23).

Performing the integrations in equation (12.23) they obtained:

$$\Phi_{ab}{}^{\text{weak}} = -\frac{4\pi}{3}N_e(\mathbf{r}_a - \mathbf{r}_b{}^*)^2\left[\int_0^\infty dv\frac{f(v)}{v}\right]$$
$$\times\left[\log\frac{\sin\phi_{\max}}{\sin\phi_{\min}} + \frac{mv^2\beta}{(Z-1)^2}\left(\frac{\pi^2}{4} - 1\right)\right]. \tag{12.24}$$

Here and below the expression $(\mathbf{r}_a - \mathbf{r}_b{}^*)^2$ stands for the scalar product (also known as the dot-product) of the operator $(\mathbf{r}_a - \mathbf{r}_b{}^*)$ with itself. In the theory of the dynamical Stark broadening of spectral lines in plasmas by electrons, the corresponding matrix elements are calculated with respect to the unperturbed wave functions.

Then the authors of paper [6] added the CT estimate for the contribution of strong collisions

$$\Phi_{ab}{}^{\text{strong}} \approx \pi v N_e \rho_{\text{We}}{}^2. \tag{12.25}$$

where ρ_{We} corresponds to $\phi_{\max}$. Expressions for $\phi_{\max}$ and $\phi_{\min}$ are given in paper [1] (in equations (9) and (10a)) as follows

$$\sin\phi_{\max} = \sqrt{\frac{3}{2}}\frac{Z(Z-1)}{\left(n_a^2 - n_b^2\right)mv}, \tag{12.26}$$

$$\sin\phi_{\min} = \frac{\dfrac{Z-1}{mv^2\rho_{\text{D}}}}{\sqrt{1 + \dfrac{(Z-1)^2}{m^2v^4\rho_{\text{D}}^2}}} \tag{12.27}$$

It should be emphasized that the factor $(n_a^2 - n_b^2)$ in the denominator of the right side of equation (12.26) was an approximate allowance by the authors of paper [1] for the contribution of the lower level b while estimating the operator $(\boldsymbol{r}_a - \boldsymbol{r}_b{}^*)$ for hydrogenic lines of spectral series other than the Lyman lines. However, for the

Lyman lines the lower (ground) level does not contribute to electron broadening operator, so that for the Lyman lines equation (12.26) should be simplified as follows:

$$\sin\phi_{\max} = \sqrt{\frac{3}{2}}\frac{Z(Z-1)}{n_a^2 mv}. \tag{12.28}$$

At relatively small velocities of perturbing electrons, the right side of equation (12.26) or equation (12.28) could exceed unity. In this case one should set $\sin\phi_{\max} = 1$, which corresponds to $\rho_{\min} = 0$, so that there would be no contribution from strong collisions. Typically, the range of such small velocities has a very low statistical weight in the electron velocity distribution.

After substituting the above formulas for $\sin\phi_{\max}$ and $\sin\phi_{\min}$ into equation (12.23), and combining the contributions from weak and strong collisions, the authors of paper [6] obtained the final results for the electron broadening operator:

$$\begin{aligned}\Phi_{ab}(\beta) = -\frac{4\pi}{3}N_e(\mathbf{r}_a - \mathbf{r}_b{}^*)^2\left[\int_0^\infty \mathrm{d}v\frac{f(v)}{v}\right]\Bigg\{\frac{1}{2}\left[1 - \frac{3}{2}\frac{Z^2(Z-1)^2}{\left(n_a^2 - n_b^2\right)^2 m^2v^2}\right]\\ + \log\left[\sqrt{\frac{3}{2}}\frac{Zv\rho_D}{\left(n_a^2 - n_b^2\right)}\sqrt{1 + \left(\frac{Z-1}{mv^2\rho_D}\right)^2}\right] + \frac{mv^2\beta}{(Z-1)^2}\left(\frac{\pi^2}{4} - 1\right)\Bigg\}\end{aligned} \tag{12.29}$$

for the non-Lyman lines and

$$\begin{aligned}\Phi_{ab}(\beta) = -\frac{4\pi}{3}N_e(\mathbf{r}_a - \mathbf{r}_b{}^*)^2\left[\int_0^\infty \mathrm{d}v\frac{f(v)}{v}\right]\Bigg\{\frac{1}{2}\left[1 - \frac{3}{2}\frac{Z^2(Z-1)^2}{n_a^4 m^2v^2}\right]\\ + \log\left[\sqrt{\frac{3}{2}}\frac{Zv\rho_D}{n_a^2}\sqrt{1 + \left(\frac{Z-1}{mv^2\rho_D}\right)^2}\right] + \frac{mv^2\beta}{(Z-1)^2}\left(\frac{\pi^2}{4} - 1\right)\Bigg\}\end{aligned} \tag{12.30}$$

for the Lyman lines. Here and below log [...] stands for the natural logarithm.

In order to determine the significance of this effect, it is necessary to evaluate the ratio

ratio

$$= \frac{\frac{3}{2}\frac{mv^2(n_aq_a - n_bq_b)}{(Z-1)^2}\left(\frac{\pi^2}{4} - 1\right)}{\frac{1}{2}\left[1 - \frac{3}{2}\frac{Z^2(Z-1)^2}{\left(n_a^2 - n_b^2\right)^2 m^2v^2}\right] + \log\left[\sqrt{\frac{3}{2}}\frac{Zv\rho_D}{\left(n_a^2 - n_b^2\right)}\sqrt{1 + \left(\frac{Z-1}{mv^2\rho_D}\right)^2}\right]} \tag{12.31}$$

for the non-Lyman lines or the ratio

$$\text{ratio} = \frac{\frac{3}{2}\frac{mv^2 n_a q_a}{(Z-1)^2}\left(\frac{\pi^2}{4} - 1\right)}{\frac{1}{2}\left[1 - \frac{3}{2}\frac{Z^2(Z-1)^2}{n_a^4 m^2 v^2}\right] + \log\left[\sqrt{\frac{3}{2}}\frac{Zv\rho_\mathrm{D}}{n_a^2}\sqrt{1 + \left(\frac{Z-1}{mv^2\rho_\mathrm{D}}\right)^2}\right]} \tag{12.32}$$

for the Lyman lines.

The authors of paper [6] presented numerical examples for several Lyman lines. As is customary in the Stark broadening theory, instead of the integration over velocities, for the numerical examples they used the mean thermal velocity v_T of the perturbing electrons. In atomic units, the mean thermal velocity v_T, the Debye radius ρ_D, and the reduced mass can be expressed as follows

$$v_\mathrm{T} = 0.1917\sqrt{\frac{T(\mathrm{eV})}{\mathrm{m}}}\rho_\mathrm{D} = 1.404 \times 10^{11}\sqrt{\frac{T(\mathrm{eV})}{N_e(\mathrm{cm}^{-3})}}m = \frac{1 + \frac{m_e}{Am_p}}{1 + \frac{2m_e}{Am_p}}, \tag{12.33}$$

where m_e is the electron mass, m_p is the proton mass, and A is the atomic number of the radiating ion ($A \approx 2Z$).

Table 12.1 presents the values of the ratio from equation (12.33) for several Lyman lines of He II at the temperature $T = 8$ eV and the electron density $N_e = 2 \times 10^{17}$ cm^{-3}.

Figure 12.1 shows the ratio from equation (12.33) versus the electron density N_e for the Stark components of the electric quantum number $|q| = 1$ of Lyman-alpha ($n = 2$), Lyman-beta ($n = 3$), and Lyman-gamma ($n = 4$) lines of He II at the temperature $T = 8$ eV.

It is seen that for the electron broadening of the Lyman lines of He II, the allowance for the effect under consideration indeed becomes significant already at electron densities $N_e \sim 10^{17}$ cm^{-3} and increases with the growth of the electron density. It should be noted that when the ratio, formally calculated by equation (12.33), becomes comparable to unity, this is the indication that the approximate

Table 12.1. Ratio from equation (12.33) for the Stark components of several Lyman lines of He II at the temperature $T = 8$ eV and the electron density $N_e = 2 \times 10^{17}$ cm^{-3}. Reproduced with permission from [6]. Copyright 2018 P Sanders and E Oks.

N	$\|q\|$	Ratio
2	1	0.3261
3	1	0.3748
3	2	0.7496
4	1	0.5156
4	2	1.0311
4	3	1.5467

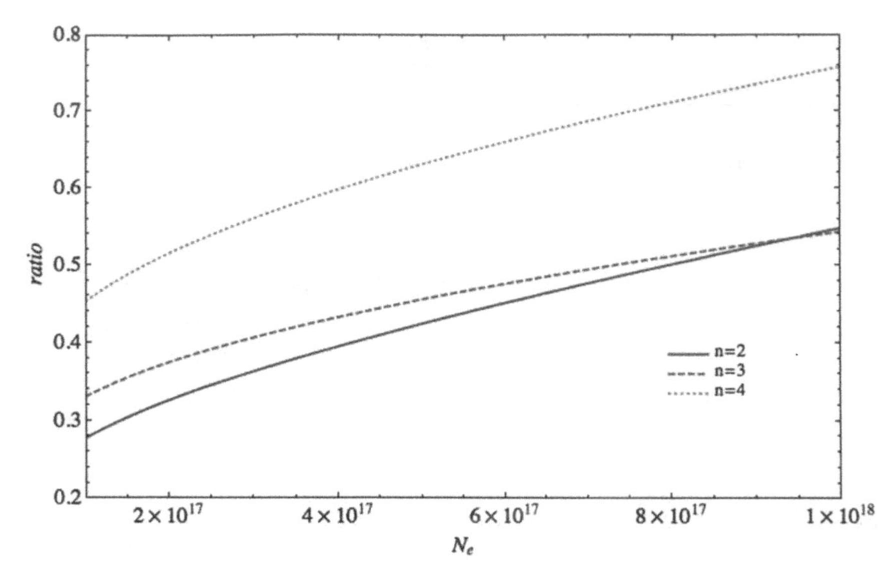

Figure 12.1. Ratio from equation (12.33) versus the electron density N_e for the Stark components of the electric quantum number $|q| = 1$ of Lyman-alpha ($n = 2$), Lyman-beta ($n = 3$), and Lyman-gamma ($n = 4$) lines of He II at the temperature $T = 8$ eV. (Reproduced with permission from [6]. Copyright 2018 P Sanders and E Oks.)

analytical treatment based on expanding equation (12.17) up to the first order of parameter β, is no longer valid. In this case the calculations should be based on solving equation (12.17) with respect to ρ without such approximation.

In summary, the authors of paper [6] considered the electron broadening of hydrogenlike spectral lines in plasmas more accurately than in the CT. In distinction to the CT, they treated it as a three-body problem involving the perturbing electron, the nucleus, and the bound electron. They employed the standard analytical method of separating rapid and slow subsystems by using the fact that the characteristic frequency of the motion of the bound electron around the nucleus is much higher than the characteristic frequency of the motion of the perturbing electron around the radiating ion.

With the help of this method they obtained more accurate analytical results for the electron broadening operator compared to the CT. By examples of the electron broadening of the Lyman lines of He II, they demonstrated that the allowance for this effect becomes significant at electron densities $N_e \sim 10^{17}$ cm^{-3} and very significant at higher densities. It is well-known that for relatively low-Z radiators, the broadening by electrons is comparable to the broadening by ions, so that the correction to the broadening by electrons, introduced in paper [6], should be significant for the total Stark width.

It is important to emphasize that the authors of paper [6] were able to obtain the above analytical results primarily due to the underlying fundamental symmetry of the class of potentials $V(R) = -A/R + B/R^2$, where A and B are constants. Namely, this class of potentials possesses an additional conserved quantity $M_{\text{eff}}^2 = M^2 + 2mB$, where M is the angular momentum and m is the mass of a particle, so that M_{eff} is the effective angular momentum. As for the impact approximation, it was not crucial to their work—they used it only for the following two purposes: first, to get the message across in a simple form, and second, for the comparison with the CT (in which the impact approximation was crucial), so that they would compare 'apples to apples' rather than 'apples to oranges'.

The authors of paper [6] mentioned that in 1981, Baryshnikov and Lisitsa [9] published very interesting results for the electron broadening of hydrogen-like spectral lines in plasmas (also presented later in book [10]) in frames of the quantum theory of the dynamical Stark broadening, while in paper [6] the results were obtained in frames of the semiclassical theory of the dynamical Stark broadening, just as in the CT. (For clarity: in the semiclassical theory, the radiating atom/ion is treated quantally, while perturbing electrons classically; in the quantum theory both the radiating atom/ion and perturbing electrons are treated quantally.) Both in paper [9] and in paper [6], they used the underlying symmetry of the class of potentials $V(R) = -A/R + B/R^2$ for obtaining analytical solutions.

A specific result for the line width Baryshnikov and Lisitsa [9] obtained for Lyman lines in the classical limit using the impact approximation, as presented in their equations (4.5) and (4.6). The authors of paper [6] compared Baryshnikov–Lisitsa's results from their equations (4.5) and (4.6) with the CT [1] for He II Lyman lines. It turned out that for $N_e \sim (10^{17}–10^{18})\ \mathrm{cm}^{-3}$, i.e. for the range of electron densities, in which the overwhelming majority of measurements of the width of He II lines were performed, Baryshnikov–Lisitsa's line width exceeds the CT line width by two orders of magnitude or more. In view of the fact that the width of He II lines, measured by various authors in benchmark experiments (i.e., experiments where plasma parameters were measured independently of the line widths), never exceeded the CT width by more than a factor of two (see, e.g., benchmark experiments [11–13]), this seems to indicate that something might be incorrect in equations (4.5) and (4.6) from paper [9] (though methodologically it was a very interesting paper). In distinction, the corrections to the CT introduced paper [6], do not exceed the factor of two for He II lines in the range of $N_e \sim (10^{17}–10^{18})\ \mathrm{cm}^{-3}$.

References

[1] Griem H R and Shen K Y 1961 *Phys. Rev.* **122** 1490

[2] Griem H R 1964 *Plasma Spectroscopy* (New York: McGraw-Hill)

[3] Griem H R 1974 *Spectral Line Broadening by Plasmas* (Cambridge, MA: Academic)

[4] Oks E 2006 *Stark Broadening of Hydrogen and Hydrogenlike Spectral Lines in Plasmas: The Physical Insight* (Oxford: Alpha Science International)

[5] Oks E 2017 *Diagnostics of Laboratory and Astrophysical Plasmas Using Spectral Lines of One-, Two-, and Three-Electron Systems* (Singapore: World Scientific)

[6] Sanders P and Oks E 2018 *J. Phys. Comm.* **2** 035033

[7] Galitski V, Karnakov B, Kogan V and Galitski V Jr 2013 *Exploring Quantum Mechanics* (Oxford: Oxford University Press) problem 8.55

[8] Kotkin G L and Serbo V G 1971 *Collection of Problems in Classical Mechanics* (Oxford: Pergamon) problem 2.3

[9] Baryshnikov F F and Lisitsa V S 1981 *Sov. Phys. JETP* **53** 471

[10] Bureyeva L A and Lisitsa V S 2000 *A Perturbed Atom Astrophysics and Space Physics Reviews* (Boca Raton, FL: CRC Press)

[11] Grützmacher K and Johannsen U 1993 *Spectral Line Shapes* vol 7 ed R Stamm and B Talin (New York: Nova Science), p 139

[12] Ahmad R 1999 *Eur. Phys. J.* D **7** 123

[13] Wrubel T, Büscher S, Kunze H-J and Ferri S 2001 *J. Phys.* B **34** 461

IOP Publishing

Analytical Advances in Quantum and Celestial Mechanics
Separating rapid and slow subsystems
Eugene Oks

Chapter 13

Concluding remarks

Separating rapid and slow subsystems is one of the most powerful analytical methods in theoretical physics. It can take various embodiments mentioned in chapter 1, such as, e.g., the 'method of averaging' [1, 2], the 'method of the effective potential' [3–5], and so on. Nevertheless, all the variations have the same thing in common: the separation of rapid and slow subsystems.

The current book assembled under one 'roof' modern applications of this method in quantum mechanics, as well as in classical mechanics in general and celestial mechanics in particular. Concerning applications to quantum mechanics, the book covered from hydrogen atoms in a high-frequency laser field to the quantum rotator-dipole in a high-frequency field, to the dynamical Stark broadening of spectral lines in plasmas—plus a formalism for the general analytical treatment of quantum systems in a high-frequency field (appendix A).

As for applications to classical mechanics in general and celestial mechanics in particular, the book presented several particular analytical solutions for the unrestricted three-body problem of celestial mechanics, as well as the classical treatment of one-electron Rydberg quasimolecules in various fields and Rydberg states of muonic-electronic helium atoms or helium-like ions—plus a generalization of the method of the effective potentials (the generalization developed in paper [6]).

We hope that the book would motivate theoretical physicists to seek, find, and implement further applications of this powerful analytical method to other quantum and classical systems. In this way, a better physical insight can be developed into complicated systems both in the micro-world (atoms, molecules and so on) and in the macro-world (the Universe).

References

[1] Bogoliubov N N and Mitropolski Y A 1961 *Asymptotic Methods in the Theory of Nonlinear Oscillations* (New York: Gordon and Breach)

[2] Oliveira A R E 2017 *Adv. Hist. Stud.* **6** 40

[3] Kapitza P L 1951 *Sov. Phys. JETP* **21** 588
[4] Kapitza P L 1951 *Uspekhi Fiz. Nauk* **44** 7
[5] Landau L D and Lifshitz E M 1976 *Mechanics* (Amsterdam: Elsevier) section 30
[6] Nadezhdin B B and Oks E 1986 *Sov. Tech. Phys. Lett.* **12** 512

IOP Publishing

Analytical Advances in Quantum and Celestial Mechanics

Separating rapid and slow subsystems

Eugene Oks

Appendix A

General analytical treatment of quantum systems in a high-frequency field

Let us consider a quantum system in a monochromatic, high-frequency field. It is described by the Hamiltonian

$$H(\mathbf{x}) = H_0(\mathbf{x}) + V(\mathbf{x}) \cos \omega t, \tag{A.1}$$

where $\mathbf{x}$ is the coordinate vector (such as, e.g., the radius-vector). To simplify formulas in the initial discussion, let us assume that the energy levels $E_k^{(0)}$ of the unperturbed Hamiltonian $H_0(\mathbf{x})$ are non-degenerate (this assumption will be relaxed later). In this situation, the quasienergies of the system contain corrections that are quadratic with respect to V: $E_k = E_k^{(0)} + E_k^{(2)}$. In the second order of the standard time-dependent perturbation theory, the expression for $E_k^{(2)}$ has the form:

$$E_k^{(2)} = [1/(2\hbar)] \sum_{\mathrm{i}} V_{ki} V_{ik} \omega_{ik} / \left(\omega^2 - \omega_{ik}^2\right), \quad \omega_{ik} = \left(E_i^{(0)} - E_k^{(0)}\right)/\hbar. \tag{A.2}$$

Here V_{ik} are the time-independent matrix elements of the perturbation and the summation is performed over all states $|i>$, including the continuum states. Formula (A.2) for $E_k^{(2)}$ is valid if $E_k^{(2)}/\hbar \ll \min(|\omega_{ik}|, \omega)$, so that, in particular, there are no resonances between the transition frequencies ω_{ik} and the field frequency ω or its harmonics.

Calculating $E_k^{(2)}$ by equation (A.2) requires summing up, a generally infinite number of terms. This makes such calculation very inefficient.

The idea behind introducing an effective potential operator U_{eff} serves the following goal: to substitute the infinite summation in equation (A.2) by calculating just one integral:

$$E_k^{(2)} = <k|U_{\text{eff}}|k>. \tag{A.3}$$

This goal can be achieved indeed for the high-frequency field, i.e., under the following condition:

$$\omega > |\omega_{ik}|. \tag{A.4}$$

For constructing U_{eff}, let us first expand the denominator in equation (A.2) in powers of $(\omega_{ik}/\omega)^2$:

$$E_k^{(2)} = \sum_i [V_{ki}V_{ik}/(2\hbar)] \sum_{s=0}^{\infty} (\omega_{ik})^{2s+1}/\omega^{2s+2}. \tag{A.5}$$

Of course, the summation over s converges under the condition (A.4). Now let us interchange the order of summations in equation (A.5):

$$E_k^{(2)} = \sum_{s=0}^{\infty}\left[\sum_i V_{ki}V_{ik}(\omega_{ik})^{2s+1}/(2\hbar\omega^{2s+2})\right]. \tag{A.6}$$

Nadezhdin [1] proved that if the sum over i in equation (A.6) converges, then this sum coincides with the average value $<k|U_s|k>$ of the following operator (below for any operators A and B, the notation $[A, B]$ is their commutator):

$$U_s = [(-1)^{s+1}/(4\hbar^{2s+2}\omega^{2s+2})]\ [[\underbrace{\ldots[V, H_0],\ldots H_0]}_{H_0\ s\text{ times}}, \underbrace{[\ldots\ [V, H_0]\ ,\ \ldots H_0]}_{H_0\ (s+1)\text{ times}}]. \tag{A.7}$$

The corresponding classical effective potentials U_s^{class} are expressed through the classical Poisson brackets {, }, rather than the commutators [,], as follows:

$$U_s^{\text{class}} = [1/(4\omega^{2s+2})]\ \{\underbrace{\{\ldots\{\ V, H_0\ \},\ \ldots H_0\}}_{H_0\ s\text{ times}},\ \underbrace{\{\ldots\{V, H_0\},\ \ldots H_0\}}_{H_0\ (s+1)\text{ times}}\}. \tag{A.8}$$

Here is one of the examples, where it is necessary to use U_s with $s > 0$. let us consider a hydrogen atom in a uniform electric field $\mathbf{F}\cos\omega t$ under the condition

$$\omega \gg \Omega_n = m_e e^4/(n\hbar)^3, \tag{A.9}$$

so that $\mathbf{F}\cos\omega t$ is the high-frequency field. (In equation (A.9), n is the principal quantum number.) The unperturbed Hamiltonian in Cartesian coordinates can be written as

$$H_0 = -[\hbar^2/(2m_e)](\partial^2/\partial x_j\partial x_j) - e^2/r. \tag{A.10}$$

Here and below the summation over suffix repeated twice is understood—in this case, over the suffix j ($j = 1, 2, 3$). The interaction potential can be represented in the form

$$V(x_j)\cos\omega t = -eF_jx_j\cos\omega t. \tag{A.11}$$

By using equation (A.7) with $s = 0$, we find:

$$U_0 = e^2F^2/(4m_e\omega^2). \tag{A.12}$$

The above quantity U_0 coincides with Kapitza's effective potential (details on Kapitza's effective potential can be found in [2–4]). From equation (A.12), it is seen that in this case Kapitza's effective potential does not depend on coordinates and therefore cannot affect the atomic energy levels.

So, it is necessary to calculate the effective potential U_1. According to equation (A.7), we have

$$U_1 = [1/(4\hbar^4\omega^4)]\ [[V, H_0], [[V, H_0], H_0]]. \tag{A.13}$$

The result of the calculation is as follows:

$$U_1 = [1/(4\hbar^4\omega^4)]F_jF_k(-e^2\partial^2 r^{-1}/\partial x_j\partial x_k), \quad r^{-1} = (x_m x_m)^{-1/2}. \tag{A.14}$$

By using equation (A.14) in the spherical polar coordinates (r, θ, φ) with the polar axis parallel to vector **F**, we find that the motion of the atomic electron occurs in the following effective potential:

$$U_{\text{eff}} = -e^2/r - \gamma(3\cos^2\theta - 1)/r^3, \quad \gamma = e^2F^2/(4m_e^2\omega^4). \tag{A.15}$$

As for U_s with $s > 1$, Nadezhdin [1] showed that for a hydrogen atom in a uniform, high-frequency electric field $\mathbf{F}\cos\omega t$, the sum over i in equation (A.6) converges as long as s does not exceed

$$s_{\max}(l) = 5/4 + l/2. \tag{A.16}$$

We recall that if the sum over i in equation (A.6) converges, then it coincides with the corresponding value of U_s. For $s > s_{\max}$, the sum over i in equation (A.6) diverges because of the contribution of the continuous spectrum.

Now let us present Nadezhdin's [1] generalization of the formalism of effective potentials to the situation where the interaction of a quantum system with a field has the form $V\cos\omega t + W\sin\omega t$. The most important example is the interaction of an atom with an elliptically-polarized electromagnetic (e.g., laser) field. In this case, the energy shift quadratic with respect to the field is

$$E_k^{(2)} = [1/(2\hbar)]\sum_m\Big[(V_{km}V_{mk} + W_{km}W_{mk})\omega_{mk}\Big/\left(\omega^2 - \omega_{mk}{}^2\right) \\ +i(V_{km}V_{mk} - W_{km}W_{mk})\omega\Big/\left(\omega^2 - \omega_{mk}{}^2\right). \tag{A.17}$$

It turns out to be possible to represent $E_k^{(2)}$ in the form

$$E_k^{(2)} = \sum_s <k|U_s|k>, \tag{A.18}$$

where

$$U_s = [(-1)^{s+1}/(4\hbar^{2s+2}\omega^{2s+2})]\{[\underbrace{[\ldots[V, H_0], \ldots H_0]}_{H_0\ s\ \text{times}}, \underbrace{[\ldots[V, H_0],\ldots H_0]}_{H_0\ (s+1)\ \text{times}}]$$
$$+ [\underbrace{[\ldots[W, H_0], \ldots H_0]}_{H_0\ s\ \text{times}}, \underbrace{[\ldots[W, H_0],\ldots H_0]}_{H_0\ (s+1)\ \text{times}}]\} \quad (A.19)$$
$$+ i[(-1)^s/(2\hbar^{2s+1}\omega^{2s+1})][\underbrace{[\ldots[V, H_0], \ldots H_0]}_{H_0\ s\ \text{times}}, \underbrace{[\ldots[W, H_0],\ldots H_0]}_{H_0\ s\ \text{times}}].$$

In particular, for Kapitza's effective potential, the corresponding generalization has the form

$$U_0 = [-1/(4\hbar^4\omega^4)]\{[V, [V, H_0]] + [W, [W, H_0]]\} + [i/(2\hbar\omega)][V, W]. \quad (A.20)$$

Now let us remove the assumption made at the beginning of appendix A that the energy levels of the unperturbed system are non-degenerate. As in the standard time-independent perturbation theory for degenerate states, the task is to find correct eigenfunctions of the zeroth order $|k\alpha>$ and then the energy corrections $E_{k\alpha}^{(2)}$. In the field $V\cos\,\omega t + W\sin\,\omega t$, the correct states of the zeroth order will be those that diagonalize the following matrix $V_{k\alpha,k\gamma}$ built on all states $|k\alpha>$, $|k\gamma>$ of the unperturbed energy $E_k^{(0)}$:

$$V_{k\alpha,k\gamma} = [1/(2\hbar)]\sum_{m,\beta}\Big\{\big(V_{k\alpha,m\beta}V_{m\beta,k\gamma} + W_{k\alpha,m\beta}W_{m\beta,k\gamma}\big)\omega_{mk}\Big/\big(\omega^2 - \omega_{mk}{}^2\big)$$
$$+ i\big(V_{k\alpha,m\beta}W_{m\beta,k\gamma} - U_{k\alpha,m\beta}W_{m\beta,k\gamma}\big)\omega\Big/\big(\omega^2 - \omega_{mk}{}^2\big)\Big\}, \quad (A.21)$$

where, e.g., $V_{k\alpha,m\beta} = <k\alpha|V|m\beta>$. Accordingly, while approximating $E_k^{(2)}$ by the effective potentials from formula (A.19), the operators U_s from (A.19) should be diagonal, i.e., in the basis of the correct functions of the zeroth order $|k\alpha>$, $|k\gamma>$, the matrix elements $<k\alpha|U_s|k\gamma>$ should be proportional to $\delta_{\alpha\gamma}$ (the latter being the Kroneker delta). The shifts of the energies $E_k^{(2)}$ are then represented by the diagonal elements of the matrix $V_{k\alpha,k\gamma}$ from (A.19) or, respectively, by the matrix elements $<k\alpha|U_s|k\alpha>$.

Finally, let us address the issue of whether the effective potentials are gauge invariant. Following Nadezhdin [1], we limit ourselves by one-dimensional systems. The interaction of a quantum system (such as, e.g., an electron in a time-independent potential) with a monochromatic field $F\cos\omega t$ can be written in the following two representations V and W

$$V = -eFx\cos\omega t,$$
$$W = (1 - \sin 2\omega t)e^2A^2/(4\,mc^2) + i(\sin\omega t)[e\hbar/(mc)]\mathrm{d}/\mathrm{d}x, \quad (A.22)$$
$$A = cF/\omega.$$

In gauge V, equation (A.2 takes the form

$$E_{k,V}{}^{(2)} = \sum_{s=0}^{\infty} <k|V_s|k>, \tag{A.23}$$

where

$$V_s = [(-1)^{s+1}e^2F^2/(4\hbar^{2s+2}\omega^{2s+2})] \ [[\underbrace{\ldots[x, H_0], \ldots H_0]}_{H_0 \ s \text{ times}}, \underbrace{[\ldots[x, H_0],\ldots H_0]}_{H_0 \ (s+1) \text{ times}}]. \tag{A.24}$$

In gauge W, equation (A.2) takes the form

$$E_{k,W}{}^{(2)} = \sum_{s=0}^{\infty} <k|W_s|k> + <k|e^2A^2/(4mc^2)|k>, \tag{A.25}$$

where

$$\begin{aligned} W_s = {} & [(-1)^{s+1}e^2A^2\hbar^2/(4m^2c^2\hbar^{2s+2}\omega^{2s+2})] \\ & [[\underbrace{\ldots[i\mathrm{d}/\mathrm{d}x, H_0], \ldots H_0]}_{H_0 \ s \text{ times}}, \underbrace{[\ldots[i\mathrm{d}/\mathrm{d}x, H_0], \ldots H_0]}_{H_0 \ (s+1) \text{ times}}]. \end{aligned} \tag{A.26}$$

Let us prove that $E_{k,V}{}^{(2)} = E_{k,W}{}^{(2)}$, i.e., the energy shift is gauge invariant, and that

$$<k|V_0|k> = <k|e^2A^2/(4mc^2)|k>, \tag{A.27}$$

$$<k|V_{s+1}|k> = <k|W_s|k>, \quad s \geqslant 0. \tag{A.28}$$

Indeed, $<k|V_0|k> = <k|e^2F^2/(4m\omega^2)|k> = <k|e^2A^2/(4mc^2)|k>$, since $A = cF/\omega$. As for proving equation (A.28), it is sufficient to show that for $s \geqslant 0$ one has

$$\begin{aligned} & [[\underbrace{\ldots[i\mathrm{d}/\mathrm{d}x, H_0], \ldots H_0]}_{H_0 \ s \text{ times}}, \underbrace{[\ldots[i\mathrm{d}/\mathrm{d}x, H_0], \ldots H_0]}_{H_0 \ (s+1) \text{ times}}] \\ & = (m^2/\hbar^4) \ [[\underbrace{\ldots[x, H_0], \ldots H_0]}_{H_0 \ (s+1) \text{ times}}, \underbrace{[\ldots[x, H_0], \ldots H_0]}_{H_0 \ (s+2) \text{ times}}]. \end{aligned} \tag{A.29}$$

Indeed:

$$[x, H_0] = [x, -(\hbar^2/(2m)\mathrm{d}^2/\mathrm{d}x^2 + U(x)] = (\hbar^2/m)\mathrm{d}/\mathrm{d}x, \tag{A.30}$$

where $U(x)$ is the potential energy. So, equations (A.28) and (A.29) are the consequences of equation (A.30).

Equation (A.28) is quite interesting. It shows that in different gauges the operators U_s with the same s have different physical meanings. For example, for the above one-dimensional quantum system, Kapitza's effective potential in gauge V is a constant, coordinate-independent value given by equation (A.27), while in gauge W Kapitza's effective potential is some function of coordinate—according to equation (A.28).

Thus, Kapitza's effective potential is not gauge invariant. A similar situation could occur in other physical problems.

References

[1] Nadezhdin B B 1986 *Radiatsionnye i Relativistskie Effekty v Atomakh i Ionakh (Radiative and Relativistic Effects in Atoms and Ions)* (Moscow: Scientific Council of the USSR Academy of Sciences on Spectroscopy), 222 in Russian

[2] Kapitza P L 1951 *Sov. Phys. JETP* **21** 588

[3] Kapitza P L 1951 *Uspekhi Fiz. Nauk* **44** 7

[4] Landau L D and Lifshitz E M 1976 *Mechanics* (Amsterdam: Elsevier) section 30

Appendix B

Analytical solution for Rydberg states of muonic-electronic negative hydrogen ion

In papers [1, 2] the authors combined two lines of research: studies of negative hydrogen ion and studies of muonic atoms/molecules (the results of these papers are also presented in appendix A of book [3]). Namely, they considered a muonic negative hydrogen ion, i.e. μpe-system and studied the possibility of circular states in such a system. They showed that the muonic motion can represent a rapid subsystem, while the electronic motion is a slow subsystem.

As the first step, the authors of papers [1, 2] obtained analytically classical energy terms (the meaning of classical energy terms is explained below) for the rapid subsystem at the frozen slow subsystem, i.e., for the quasimolecule where the muon rotates around the axis connecting the immobile proton and the immobile electron, and found that the muonic motion is stable.

Then the authors of papers [1, 2] unfroze the slow subsystem and studied a slow revolution of the axis connecting the proton and electron. They derived the condition, under which the separation into the rapid and slow subsystems is valid.

Finally, the authors of papers [1, 2] demonstrated that the spectral lines, emitted by the muon in the quasimolecule μpe, are red-shifted compared to the corresponding spectral lines that would have been emitted by the muon in a muonic hydrogen atom (in the μp-subsystem). The experimental observation of this red shift should be one of the possibilities of detecting the formation of such muonic negative hydrogen ions.

Below we present details following paper [1].

So, we consider a hydrogen atom with a muon rotating in a circle perpendicular to and centered at the axis connecting the proton and the electron—see figure B.1. As we show below, in this configuration the muon may be considered the rapid subsystem while the proton and electron will be the slow subsystem, which

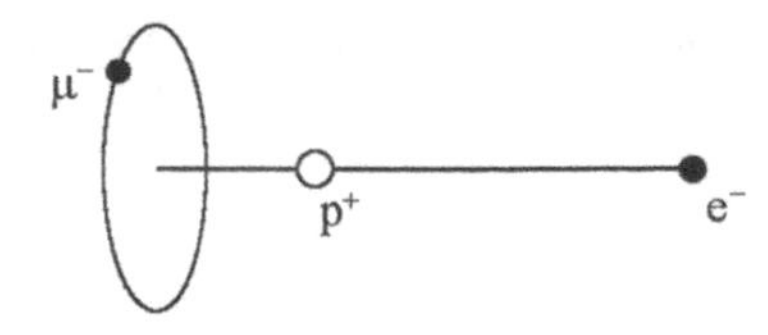

Figure B.1. A muon rotating in a circle perpendicular to and centered at the axis connecting the proton and the electron. (Reproduced with permission from [3]. Copyright 2015 World Scientific.)

essentially reduces the problem to the two stationary Coulomb center problem, where the effective stationary 'nuclei' will be the proton and electron. The straight line connecting the proton and electron will be called here 'internuclear' axis. We use the atomic units in this study.

A detailed classical analytical solution of the two-stationary-Coulomb-center problem, where an electron revolves around nuclei of charges Z and Z', has been presented in papers [4, 5]. In papers [1, 2] some results were based on the results obtained in papers [4, 5].

The Hamiltonian of the rotating muon is

$$H = \left(p_z^2 + p_\rho^2 + p_\varphi^2/\rho^2\right)/(2m) - Z/(z^2 + \rho^2)^{1/2} - Z'/[(R - z)^2 + \rho^2], \quad \text{(B.1)}$$

where m is the mass of the muon (in atomic units $m = 206.7682746$), Z and Z' are the charges of the effective nuclei (in our case, $Z = 1$ and $Z' = -1$), R is the distance between the effective nuclei, (ρ, φ, z) are the cylindrical coordinates, in which Z is at the origin and Z' is ar $z = R$, and (p_ρ, p_φ, p_z) are the corresponding momenta of the muon.

Since φ is a cyclic coordinate, the corresponding momentum is conserved:

$$|p_\varphi| = \text{const} = L. \quad \text{(B.2)}$$

With this substituted into equation (B.1), we obtain the Hamiltonian for the z- and ρ-motions

$$H_{z\rho} = \left(p_z^2 + p_\rho^2\right)/2 + U_{\text{eff}}(z, \rho), \quad \text{(B.3)}$$

where an effective potential energy is:

$$U_{\text{eff}}(z, \rho) = L^2/(2m\rho^2) - Z/(z^2 + \rho^2)^{1/2} - Z'/[(R - z) + \rho^2]. \quad \text{(B.4)}$$

Since in a circular state $p_z = p_\rho = 0$, the total energy $E(z, \rho) = U_{\text{eff}}(z, \rho)$.

With $Z = 1$, $Z' = -1$ and the scaled quantities

$$w = z/R, \quad v = \rho/R, \quad \varepsilon = -ER, \quad \ell = L/(mR)^{1/2}, \quad r = mR/L^2, \quad \text{(B.5)}$$

we obtain the scaled energy ε of the muon:

$$\varepsilon = 1/(w^2 + v^2)^{1/2} - 1/[(1 - w)^2 + v^2]^{1/2} - \ell^2/(2v^2) \quad \text{(B.6)}$$

The equilibrium condition with respect to the scaled coordinate w is $\partial\varepsilon/\partial w = 0$; the result can be brought to the form:

$$[(1 - w)^2 + v^2]^{3/2}/(w^2 + v^2)^{3/2} = (w - 1)/w. \tag{B.7}$$

Since the left side of equation (B.7) is positive, the right side must be also positive: $(w - 1)/w > 0$. Consequently, the allowed ranges of w here are $-\infty < w < 0$ and $1 < w < +\infty$. This means that equilibrium positions of the center of the muon orbit could exist (judging only by the equilibrium with respect to w) either beyond the proton or beyond the electron, but there are no equilibrium positions between the proton and electron.

Solving equation (B.7) for v^2 and denoting $v^2 = p$, we obtain:

$$p(w) = w^{2/3}(w - 1)^{2/3}[w^{2/3} + (w - 1)^{2/3}]. \tag{B.8}$$

The equilibrium condition with respect to the scaled coordinate v is $\partial\varepsilon/\partial v = 0$, which yields:

$$\ell^2 = p^2\{1/(w^2 + p)^{3/2} - 1/[(1 - w)^2 + p]^{3/2}\}. \tag{B.9}$$

Since the left side of equation (B.9) is positive, the right side must be also positive. This entails the relation $w^2 + p < (1 - w)^2 + p$, which simplifies to $2w - 1 < 0$, which requires $w < 1/2$.

Thus, the equilibrium with respect to both w and v is possible only in the range $-\infty < w < 0$, while in the second range, $1 < w < +\infty$ (derived from the equilibrium with respect to w only) there is no equilibrium with respect to v.

From the last two relations in equation (B.5), we find $r = 1/\ell^2$; thus

$$r = p^{-2}\{1/(w^2 + p)^{3/2} - 1/[(1 - w)^2 + p]^{3/2}\}^{-1}, \tag{B.10}$$

where p is given by equation (B.8). Therefore, the quantity r in equation (B.10) is the scaled 'internuclear' distance dependent on the scaled internuclear coordinate w.

Now we substitute the value of ℓ from equation (B.9), as well as the value of p from equation (B.8) into equation (B.6), obtaining $\varepsilon(w)$—the scaled energy of the muon dependent on the scaled internuclear coordinate w. Since $E = -\varepsilon/R$ and $R = rL^2/m$, then $E = -(m/L^2)\varepsilon_1$ where $\varepsilon_1 = \varepsilon/r$. The parametric dependence $\varepsilon_1(r)$ will yield the energy terms.

The form of the parametric dependence $\varepsilon_1(r)$ can be significantly simplified by introducing a new parameter $\gamma = (1 - 1/w)^{1/3}$. The region $-\infty < w < 0$ corresponds to $1 < \gamma < \infty$. The parametric dependence will then have the following form:

$$\varepsilon_1(\gamma) = (1 - \gamma)^4(1 + \gamma^2)^2/[2(1 - \gamma + \gamma^2)^2(1 + \gamma^2 + \gamma^4)], \tag{B.11}$$

$$r(\gamma) = (1 + \gamma^2 + \gamma^4)^{3/2}/[\gamma(1 + \gamma^2)^2], \tag{B.12}$$

Classical energy terms given by the parametric dependence of the scaled energy $\varepsilon_1 = (L^2/m)E$ on the scaled internuclear distance $r = (m/L^2)R$ are presented in figure B.2.

Figure B.2 actually contains two coinciding energy terms: there is a double degeneracy with respect to the sign of the projection of the muon angular

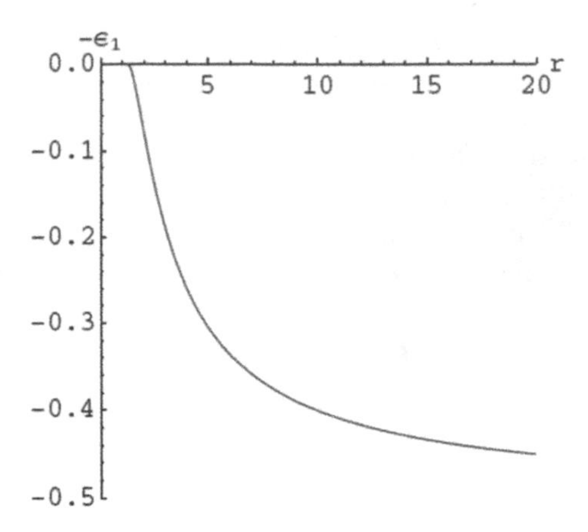

Figure B.2. Classical energy terms: the scaled energy $-\varepsilon_1 = (L^2/m)E$ versus the scaled internuclear distance $r = (m/L^2)R$. (Reproduced with permission from [3]. Copyright 2015 World Scientific.)

momentum on the internuclear axis. We remind readers that L is the absolute value of this projection—in accordance with its definition in equation (B.2).

The minimum value of R, corresponding to the point where the term starts, can be found from equation (B.12). The term starts at $w = -\infty$, which corresponds to $\gamma = 1$; taking the value of equation (B.12) at this point, we find

$$R_{\min} = (3^{3/2}/4)L^2/m. \tag{B.13}$$

With the value of $m = 206.7682746$, equation (B.13) yields $R = 0.00628258\ L^2$.

The following note might be useful. The plot in figure B.1 represents two degenerate classical energy terms of 'the same symmetry'. (In the physics of diatomic molecules, the terminology 'energy terms of the same symmetry' means the energy terms of the same projection of the angular momentum on the internuclear axis.) For a given R and L, the classical energy E takes only one *discrete* value. However, as L varies over a *continuous* set of values, so does the classical energy E (as it should be in classical physics).

The revolution frequency of the muon Ω is

$$\Omega = L/(m\rho^2) = L/(mR^2v^2) = L/(mR^2p) \tag{B.14}$$

in accordance with the previously introduced notations $p = v^2 = (\rho/R)^2$. Since $R = L^2r/m$ (see equation (B.5)), then equation (B.14) becomes $\Omega = (m/L^3)f$, where $f = 1/(pr^2)$. Using equation (B.12) for $r(\gamma)$ and equation (B.8) for $p(w)$ with the substitution $w = 1/(1 - \gamma^3)$, where $\gamma > 1$, we finally obtain:

$$\Omega = (m/L^3)f(\gamma), \quad f(\gamma) = (1 + \gamma^2)^3(1 - \gamma^3)^2/(1 + \gamma^2 + \gamma^4)^3, \tag{B.15}$$

where $f(\gamma)$ is the scaled muon revolution frequency. Figure B.3 shows the scaled muon revolution frequency $f = (L^3/m)\omega$ versus the scaled internuclear distance $r = (m/L^2)R$.

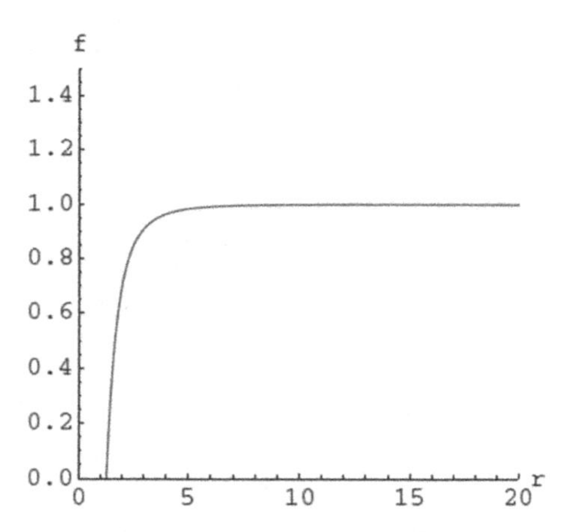

Figure B.3. The scaled muon revolution frequency $f = (L^3/m)\Omega$ versus the scaled internuclear distance $r = (m/L^2)R$. (Reproduced with permission from [3]. Copyright 2015 World Scientific.)

It is seen that for almost all values of the scaled internuclear distance $r = (m/L^2)R$, the scaled muon revolution frequency $f = (L^3/m)\Omega$ is very close to its maximum value $f_{max} = 1$, corresponding to large values of R. (The quantity f_{max} can be easily found from equation (B.15) given that large values of R correspond to $\gamma \gg 1$ and that this limit yields $f_{max} = 1$.) In other words, for almost all values of R, the muon revolution frequency Ω is very close to its maximum value

$$\Omega_{max} = m/L^3. \tag{B.16}$$

Further, below we will compare the muon revolution frequency with the corresponding frequency of the electronic motion and derive the condition of validity of the separation into rapid and slow subsystems.

For analysing the stability of the muon motion, corresponding to the degenerate classical energy terms, we use the same approach as in paper [5]. Namely, in paper [5], while considering a classical circular motion of a charged particle (which was the electron in [5]) in the field of two stationary Coulomb centers, using the same notations as in the present paper, it was shown that the frequencies of small oscillations of the scaled coordinates w and v of the circular orbit around its equilibrium position are given by

$$\omega_{\pm} = [1/(1 - w) \pm 3w/Q]^{1/2}/(w^2 + p)^{3/4} \tag{B.17}$$

where

$$Q = (w^2 + p)^{1/2}[(1 - w)^2 + p]^{1/2} \tag{B.18}$$

These oscillations are in the directions (w', v') obtained by rotating the (w, v) coordinates by the angle α:

$$\delta w' = \delta w \cos \alpha + \delta v \sin \alpha \quad \delta v' = -\delta w \sin \alpha + \delta v \cos \alpha \tag{B.19}$$

where the 'δ' symbol stands for the small deviation from equilibrium. The angle α is determined by the following relation:

$$\alpha = (1/2)\tan^{-1}\{(1-2w)p^{1/2}/[w(1-w)+p]\} \tag{B.20}$$

The quantity Q in equation (B.18) is always positive since it contains the squares of the coordinates. From equation (B.17), it is seen that the condition for both frequencies to be real is

$$1/(1-w) \geqslant 3w/Q \tag{B.21}$$

For the frequency ω_- to be real, equation (B.17) requires $Q \geqslant 3w(1-w)$. For any $w < 0$ (which is the allowed range of w), this inequality is satisfied: the left side is always positive while the right side is always negative.

For the frequency ω_+ to be real, the following function $F(w)$ must be positive (in accordance with equations (B.17) and (B.18)):

$$F(w) = (w^2+p)[(1-w)^2+p] - 9w^2(1-w)^2 \tag{B.22}$$

After replacing w by $\gamma = (1-1/w)^{1/3}$, expression (B.22) becomes

$$F(\gamma) = \gamma^2(\gamma^2-1)^2(1+4\gamma^2+\gamma^4)/(\gamma^3-1)^4 \tag{B.23}$$

Since the allowed range of $w < 0$ corresponds to $\gamma > 1$, it is seen that $F(\gamma)$ is always positive.

Thus, the corresponding classical energy terms correspond to the stable motion.

Now we discuss the electronic motion and the validity of this entire scenario. We unfreeze the slow subsystem and analyse a slow revolution of the axis connecting the proton and electron, the electron executing a circular orbit. In accordance to the concept of separating rapid and slow subsystems, the rapid subsystem (the revolving muon) follows the adiabatic evolution of the slow subsystem. This means that the slow subsystem can be treated as a modified 'rigid rotator' consisting of the electron, the proton, and the ring, over which the muon charge is uniformly distributed, all distances within the system being fixed (see figure B.1).

The potential energy of the electron in atomic units (with the angular-momentum term) is

$$E_e = M^2/(2R^2) - 1/R + 1/[\rho^2+(R-z)^2]^{1/2} \tag{B.24}$$

where M is the electronic angular momentum. Its derivative by R must vanish at equilibrium, which yields

$$dE_e/dR = -M^2/R^3 + 1/R^2 - (R-z)/[\rho^2+(R-z)^2]^{3/2} = 0 \tag{B.25}$$

which gives us the value of the scaled angular momentum

$$\ell_e = M/R^{1/2} \tag{B.26}$$

corresponding to the equilibrium:

$$\ell_e^2 = 1 - (1 - w)/[(1 - w)^2 + p]^{3/2} \tag{B.27}$$

where the scaled quantities w, p of the muon coordinates are defined in equation (B.5). Using the muon equilibrium condition from equation (B.7) with v^2 denoted as p, we can represent equation (B.27) in the form:

$$\ell_e^2 = 1 + w/(w^2 + p)^{3/2}. \tag{B.28}$$

After replacing w by $\gamma = (1 - 1/w)^{1/3}$, we obtain

$$\ell_e(\gamma) = [1 - (1 - \gamma)^2(1 + \gamma + \gamma^2)^{1/2}/(1 - \gamma + \gamma^2)^{3/2}]^{1/2} \tag{B.29}$$

The electron revolution frequency is $\omega = M/R^2 = \ell_e(\gamma)/R^{3/2}$ given that $M = \ell_e(\gamma)R^{1/2}$ in accordance with equation (B.26). Since $R = L^2r(\gamma)/m$ (see equation (B.5)) with $r(\gamma)$ given by equation (B.12), then from $\omega = \ell_e(\gamma)/R^{3/2}$ we obtain

$$\omega = m^{3/2}\ell_e(\gamma)/\{L^3[r(\gamma)]^{3/2}\}. \tag{B.30}$$

From equations (B.15) and (B.30), we find the following ratio of the muon and electron revolution frequencies:

$$\Omega/\omega = (1/m^{1/2})f(\gamma)[r(\gamma)]^{3/2}/\ell_e(\gamma), \tag{B.31}$$

where $f(\gamma)$ is given in equation (B.15).

In addition to the above relation $R = L^2r(\gamma)/m$, the same quantity R can be expressed from equation (B.26) as $R = M^2/[\ell_e(\gamma)]^2$. Equating the right sides of these two expressions, we obtain the equality $L^2r(\gamma)/m = M^2/[\ell_e(\gamma)]^2$, from which it follows:

$$L/M = m^{1/2}/\{\ell_e(\gamma)[r(\gamma)]^{1/2}\}. \tag{B.32}$$

The combination of equations (B.31) and (B.32) represent an analytical dependence of the ratio of the muon and electron revolution frequencies Ω/ω versus the ratio of the muon and electron angular momenta L/M via the parameter γ as the latter varies from 1 to ∞. This dependence is presented in figure B.4.

For the separation into the rapid and slow subsystems to be valid, the ratio of frequencies Ω/ω should be significantly greater than unity. From figure B.4, it is seen that this requires the ratio of angular momenta L/M to be noticeably greater than 20.

There is another validity condition to be checked for this scenario. Namely, the revolution frequency Ω of the muon must be also much greater than the inverse lifetime of the muon $1/T_{\text{life}}$, where $T_{\text{life}} = 2.2\ \mu\text{s} = 0.91 \times 10^{11}$ a.u.: $\Omega T_{\text{life}} \gg 1$. Since for almost all values of R, the muon revolution frequency Ω is very close to its maximum value $\Omega_{\max} = m/L^3$, as shown above, then the second validity condition can be estimated as $(m/L^3)T_{\text{life}} \gg 1$, from which it follows

$$L \ll L_{\max} = (m\ T_{\text{life}})^{1/3} = 26\,600 \tag{B.33}$$

(we remind that $m = 206.7682746$ in atomic units). So, the second validity condition is fulfilled for any practically feasible value of the muon angular momentum L.

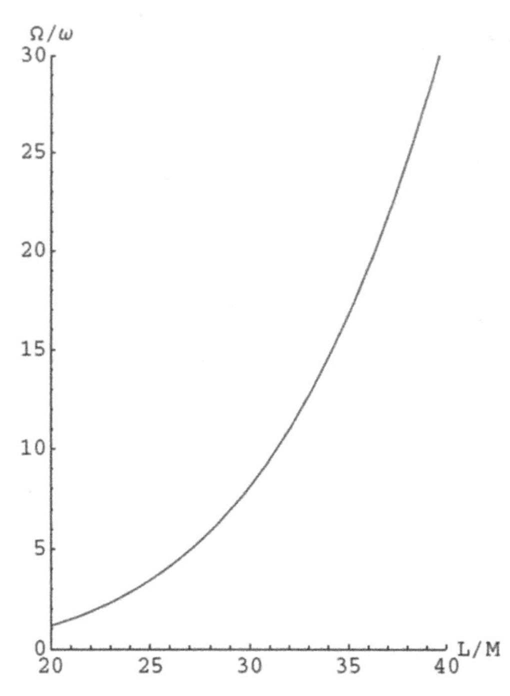

Figure B.4. The ratio of the muon and electron revolution frequencies Ω/ω versus the ratio of the muon and electron angular momenta L/M. (Reproduced with permission from [3]. Copyright 2015 World Scientific.)

Thus, for the ratio of angular momenta L/M noticeably >20, we deal here with a muonic quasimolecule where the muon rapidly rotates about the axis connecting the proton and electron following a relatively slow rotation of this axis.

The authors of paper [1] also performed simulations of the muonic and electronic motion by solving numerically the corresponding Newton's equation. The simulations confirmed their analytical results.

Now we discuss the red shift of spectral lines of the muonic–electronic negative hydrogen ions in comparison with muonic hydrogen atoms. The muon, rotating in a circular orbit at the frequency $\Omega(R)$, should emit a spectral line at this frequency. The maximum value $\Omega_{max} = m/L^3$ corresponds to the frequency of spectral lines emitted by the muonic hydrogen atom (by the μp-subsystem). For the equilibrium value of the proton–electron separation—just as for almost all values of R—the frequency Ω is slightly smaller than Ω_{max}. Therefore, the spectral lines, emitted by the muon in the quasimolecule μpe, experience a red shift compared to the corresponding spectral lines that would have been emitted by the muon in a muonic hydrogen atom. The relative red shift δ is defined as follows

$$\delta = (\lambda - \lambda_0)/\lambda_0 = (\Omega_{max} - \Omega)/\Omega, \tag{B.34}$$

where λ and λ_0 are the wavelength of the spectral lines for the quasimolecule μpe and the muonic hydrogen atom, respectively. Using equation (B.15), the relative red shift can be represented in the form

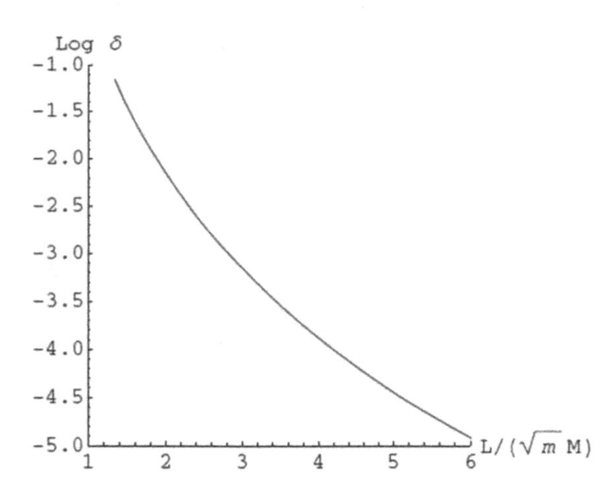

Figure B.5. Universal dependence of the relative red shift δ of the spectral lines of the quasimolecule μpe (or πpe) on $L/(m^{1/2}M)$, which is the ratio of the muon and electron angular momenta L/M divided by the square root of the mass m of the muon or pion. (Reproduced with permission from [3]. Copyright 2015 World Scientific.)

$$\delta(\gamma) = 1/f(\gamma) - 1, \tag{B.35}$$

where $f(\gamma)$ is given in equation (B.15).

The combination of equations (B.35) and (B.32) represents an analytical dependence of the relative red shift δ on the ratio of the muon and electron angular momenta L/M via the parameter γ as the latter varies from 1 to ∞. Figure B.5 presents the dependence of δ on $L/(m^{1/2}M)$. In this form the dependence is 'universal', i.e., valid for different values of the mass m: for example, it is valid also for the quasimolecule πpe where there is a pion instead of the muon. Figure B.6 presents the dependence of δ on L/M specifically for the quasimolecule μpe.

It is seen that the relative red shift of the spectral lines is well within the spectral resolution $\Delta\lambda_{res}/\lambda$ of available spectrometers: $\Delta\lambda_{res}/\lambda \sim (10^{-4}\text{–}10^{-5})$ as long as the ratio of the muon and electron angular momenta $L/M < 80$. Thus, this red shift can be observed and this would be one of the ways to detect the formation of such muonic negative hydrogen ions.

Figure B.7 presents the dependence of the relative red shift δ on the ratio of the muon and electron revolution frequencies Ω/ω. It is seen that the relative red shift decreases as the ratio of the muon and electron revolution frequencies increases, but it remains well within the spectral resolution $\Delta\lambda_{res}/\lambda$ of available spectrometers.

In summary, in papers [1, 2] the authors studied the existence of a muonic negative hydrogen ion (a 'molecule' μpe consisting of a proton, an electron and a muon) with the muon and electron being in circular states. They found out that this is indeed possible. In this case, the muonic motion can represent a rapid subsystem while the electronic motion is a slow subsystem.

The authors of papers [1, 2] demonstrated that the system has a classical double-degenerate energy term, corresponding to a stable motion. Then they studied a slow

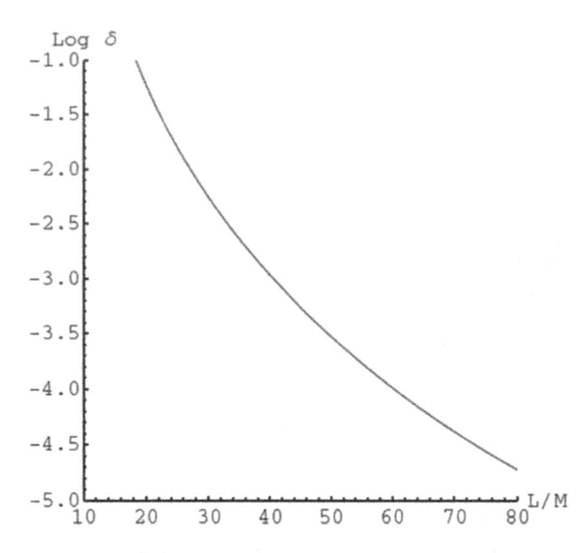

Figure B.6. Dependence of the relative red shift δ of the spectral lines of the quasimolecule μpe on the ratio of the muon and electron angular momenta L/M. (Reproduced with permission from [3]. Copyright 2015 World Scientific.)

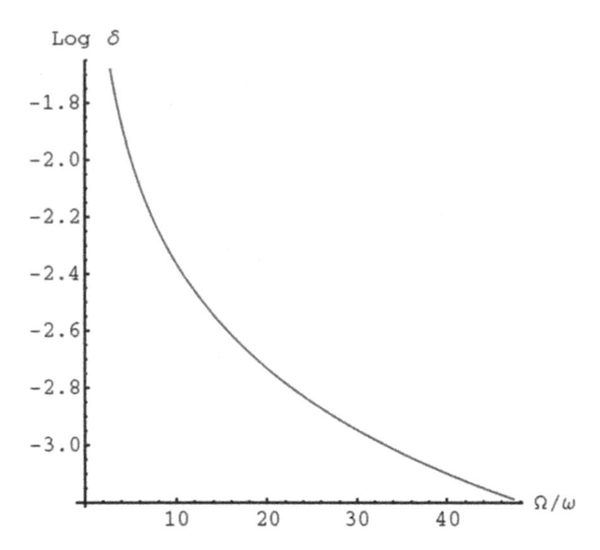

Figure B.7. Dependence of the relative red shift δ on the ratio of the muon and electron revolution frequencies Ω/ω. (Reproduced with permission from [3]. Copyright 2015 World Scientific.)

revolution of the axis connecting the proton and electron and derived the validity condition for the analytical method they used.

Finally, they found out the red shift of the spectral lines, emitted by the muon in the quasimolecule μpe, compared to the corresponding spectral lines that would have been emitted by the muon in a muonic hydrogen atom. It turned out that the relative values of this red shift, are significantly greater than the resolution of available spectrometers. Thus, the experimental observation of this red shift is possible and it would be one of the possibilities of detecting the formation of such muonic negative hydrogen ions.

References

[1] Kryukov N and Oks E 2012 *Inter. Rev. Atom. Mol. Phys.* **3** 17
[2] Kryukov N and Oks E 2013 *Can. J. Phys.* **91** 715
[3] Oks E 2015 *Breaking Paradigms in Atomic and Molecular Physics* (Singapore: World Scientific)
[4] Oks E 2000 *Phys. Rev. Lett.* **85** 2084
[5] Oks E 2000 *J. Phys. B: At. Mol. Opt. Phys.* **33** 3319

IOP Publishing

Analytical Advances in Quantum and Celestial Mechanics

Eugene Oks

Appendix C

An alternative analytical solution for Rydberg states of muonic–electronic helium-like atoms or ions

In appendix B, we followed papers [1, 2] to present the study of a system consisting of a proton, a muon, and an electron, where the muon and the electron were in circular states. In appendix C, we briefly present a generalization of that study where the proton in the μpe quasimolecule is replaced by a fully-stripped ion of a nuclear charge $Z > 1$ (quasimolecules as μZe). It turns out that just as in the previously studied case of $Z = 1$, the muonic motion can represent a rapid subsystem while the electronic motion can represent a slow subsystem. Below we present selected excerpts from paper [3] where such a study was presented.

The equation for the effective potential energy is the same as equation (B.4) from [1]

$$U_{\text{eff}}(z, \rho) = \frac{L^2}{2m\rho^2} - \frac{Z}{\sqrt{z^2 + \rho^2}} - \frac{Z'}{\sqrt{(R - z)^2 + \rho^2}} \tag{C.1}$$

where (ρ, φ, z) are the cylindrical coordinates, m is the mass of the muon (in atomic units $m = 206.768\,2746$), Z and Z' are the charges of the effective nuclei (in our case, $Z' = -1$), R is the distance between the effective nuclei. Using the scaled quantities defined by equation (B.5) from [1]

$$w = \frac{z}{R},\ v = \frac{\rho}{R},\ \varepsilon = -ER,\ \ell = \frac{L}{\sqrt{mR}},\ r = \frac{mR}{L^2} \tag{C.2}$$

one gets the following equation for the scaled energy of the muon:

$$\varepsilon = \frac{Z}{\sqrt{w^2 + p}} - \frac{1}{\sqrt{(1 - w)^2 + p}} - \frac{\ell^2}{2p} \tag{C.3}$$

where $p \equiv v^2$ is the squared scaled radial coordinate. The requirement for the derivative of the scaled energy with respect to the axial and radial coordinates (w, p) to vanish at equilibrium, yields the following two equations:

$$p = w^{2/3}(w - 1)^{2/3}\frac{Z^{2/3}(w - 1)^{4/3} - w^{4/3}}{(w - 1)^{2/3} - Z^{2/3}w^{2/3}} \tag{C.4}$$

$$\ell^2 = p^2\left(\frac{Z}{(w^2 + p)^{3/2}} - \frac{1}{((1 - w)^2 + p)^{3/2}}\right) \tag{C.5}$$

Since the left-hand sides of equations (C.4) and (C.5) are always positive, this imposes the conditions on the equilibrium range. For $Z = 1$, it was $w < 0$. In the case of $Z > 1$ it is not half-infinite—it has a lower limit:

$$-\frac{1}{Z - 1} < w < 0 \tag{C.6}$$

The analysis of (C.4) and (C.5) also shows that there are no equilibrium points for $w > 0$,

On substituting the value of ℓ^2 from equation (C.5) into equation (C.3) and then the value of p from equation (C.4) into the resulting equation, one obtains $\varepsilon(w)$—the dependence of the scaled energy of the muon on the scaled internuclear coordinate w for a given value of Z. From the scaling (C.2) one has $r = 1/\ell^2$. It is convenient to redefine the scaled energy as $\varepsilon_1 = \varepsilon/r$: it has the same scaling as r.

After defining

$$\gamma = \left(1 - \frac{1}{w}\right)^{1/3} \tag{C.7}$$

the parametric dependence ε_1 versus r (via parameter γ) yields the energy terms for a given Z:

$$\varepsilon_1 = \frac{(Z^{2/3}\gamma^4 - 1)^2(Z^{2/3}\gamma(\gamma^3 + 2) - 2\gamma^3 - 1)}{2(\gamma^3 - 1)(\gamma^3 + 1)^3} \tag{C.8}$$

$$r = \frac{\sqrt{(\gamma^6 - 1)^3(\gamma^2 - Z^{2/3})}}{\gamma(Z^{2/3}\gamma^4 - 1)^2} \tag{C.9}$$

Figure C1 spresents the energy terms for the values of $Z = 2$, 3, 4 and 5.

The behavior of the muon revolution frequency is similar to the case of $Z = 1$, but now the maximum value of it is Z^2m/L^3, which is the Kepler frequency for the muonic hydrogen-like ion of the nuclear charge Z.

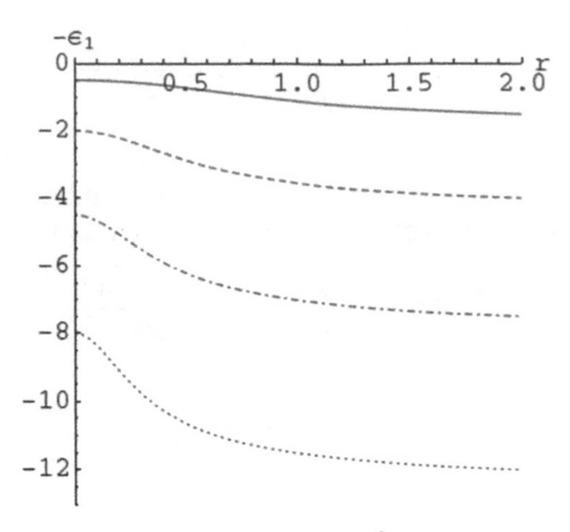

Figure C1. Classical energy terms: the scaled energy $-\varepsilon_1 = (L^2/m)E$ versus the scaled internuclear distance $r = (m/L^2)R$ for $Z = 2$ (solid curve), $Z = 3$ (dashed curve), $Z = 4$ (dot-dashed curve) and $Z = 5$ (dotted curve). (Reproduced with permission from [3]. Copyright 2014 N Kryukov and E Oks.)

The condition for the separation of the rapid and slow subsystems was that the ratio Ω/ω of the muon and electron revolution frequencies should be much greater than unity. For this to be valid, the case of $Z = 1$ required the ratio L/M of the muon and electron angular momenta to be noticeably greater than 20. Calculations show that as Z increases, the required ratio L/M increases to maintain the same condition for Ω/ω.

The equations for the frequency of the muon (Ω) and of the electron (ω) are as follows:

$$\Omega = \frac{m}{L^3}\nu,\ \nu = \frac{(Z^{2/3}\gamma^4 - 1)^3}{(\gamma^3 - 1)(\gamma^3 + 1)^3} \tag{C.10}$$

$$\omega = \frac{m^{3/2}}{L^3}\frac{\ell_e}{r^{3/2}} \tag{C.11}$$

where r is given by equation (C.9) and ℓ_e is the equilibrium value of the scaled electron angular momentum $M/R^{1/2}$ (M is the electron angular momentum in atomic units):

$$\ell_e = \sqrt{Z}\sqrt{1 - \frac{(1-\gamma)^2\sqrt{1+\gamma+\gamma^2}}{(1-\gamma+\gamma^2)^{3/2}}} \tag{C.12}$$

From these equations, the ratio of the muon's and electron's frequencies as well as the ratio of the angular momenta are as follows:

$$\frac{\Omega}{\omega} = \frac{1}{\sqrt{m}}\frac{\nu r^{3/2}}{\ell_e} \tag{C.13}$$

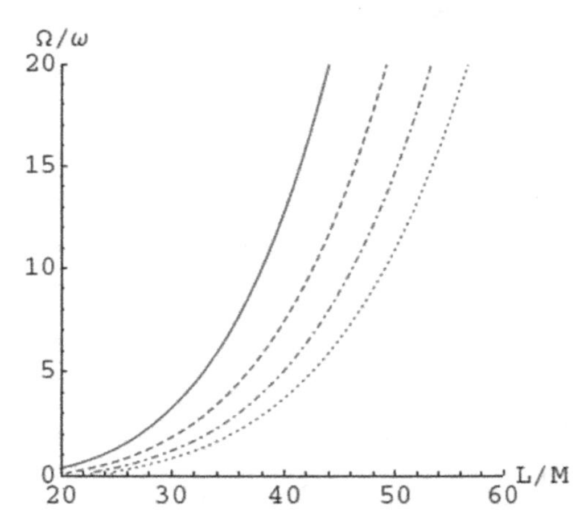

Figure C2. The ratio of the muon and electron revolution frequencies Ω/ω versus the ratio of the muon and electron angular momenta L/M for Z = 2 (solid curve), $Z = 3$ (dashed curve), $Z = 4$ (dot-dashed curve) and $Z = 5$ (dotted curve). (Reproduced with permission from [3]. Copyright 2014 N Kryukov and E Oks.)

$$\frac{L}{M} = \frac{\sqrt{m}}{\ell_e \sqrt{r}} \tag{C.14}$$

Figure C2 shows the mutual dependence of these two ratios for selected values of Z. It is seen that, for example, for $Z = 5$, the ratio L/M must be noticeably greater than 34 in order to satisfy the validity condition $\Omega/\omega \gg 1$.

The spectral lines emitted by the muon experience a red shift compared to the spectral lines of the corresponding muonic hydrogen-like ions. The relative red shift, defined as

$$\delta = \frac{\Omega_{\max} - \Omega}{\Omega} \tag{C.15}$$

is represented as

$$\delta = \frac{1}{\nu(\gamma)} - 1 \tag{C.16}$$

Figure C3 shows the dependence of the relative red shift with respect to the ratio of muon and electron angular momenta for some typical values of Z ($\log x = \log_{10} x$)—the dependence given by the parametric equations (C.16) and (C.14).

Figure C4 shows the dependence of the relative red shift on the ratio of muon and electron frequencies for the same values of Z—the dependence given by the parametric equations (C.16) and (C.13).

It is seen that the relative red shift decreases as Z increases. However, it remains within the spectral resolution of available spectrometers ($10^{-4} - 10^{-5}$) at least up to $Z = 5$.

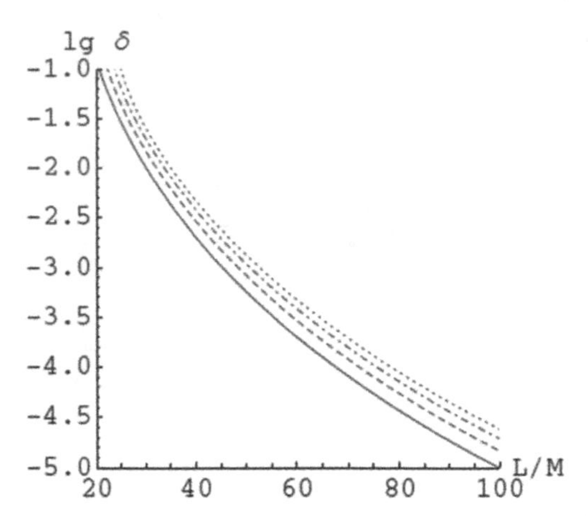

Figure C3. Dependence of the relative red shift δ of the spectral lines of the quasi-molecule μZe on the ratio of the muon and electron angular momenta for $Z = 2$ (solid curve), $Z = 3$ (dashed curve), $Z = 4$ (dot-dashed curve) and $Z = 5$ (dotted curve). (Reproduced with permission from [3]. Copyright 2014 N Kryukov and E Oks.)

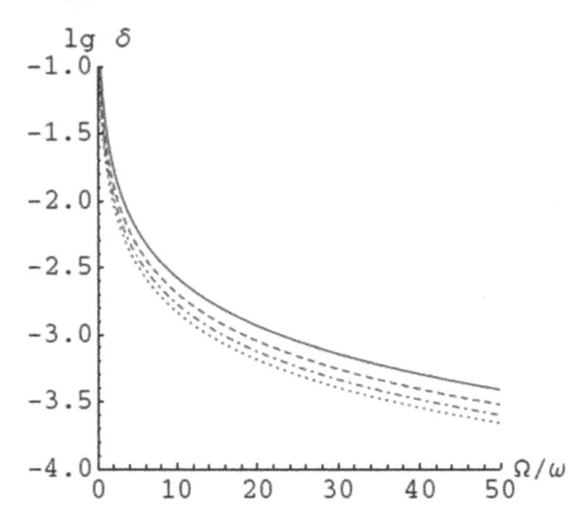

Figure C4. Dependence of the relative red shift δ of the spectral lines of the quasi-molecule μZe on the ratio of the muon and electron frequencies for $Z = 2$ (solid curve), $Z = 3$ (dashed curve), $Z = 4$ (dot-dashed curve) and $Z = 5$ (dotted curve). (Reproduced with permission from [3]. Copyright 2014 N Kryukov and E Oks.)

References

[1] Kryukov N and Oks E 2012 *Inter. Rev. Atom. Mol. Phys.* **3** 17
[2] Kryukov N and Oks E 2013 *Can. J. Phys.* **91** 715
[3] Kryukov N and Oks E 2014 *Can. J. Phys.* **92** 1405

CPSIA information can be obtained
at www.ICGtesting.com
Printed in the USA
BVHW010617090321
602058BV00003B/22